Introduction to Low Voltage Systems

Second Edition

LABORATORY MANUAL

Amy DiPaola
Samuel DiPaola

DELMAR
CENGAGE Learning

Australia • Brazil • Japan • Korea • Mexico • Singapore • Spain • United Kingdom • United States

DELMAR
CENGAGE Learning

Lab Manual to accompany
Introduction to Low Voltage Systems,
Second Edition
Amy DiPaola, Samuel DiPaola

Vice President, Editorial: Dave Garza
Director of Learning Solutions: Sandy Clark
Acquisitions Editor: Stacy Masucci
Managing Editor: Larry Main
Senior Product Manager: John Fisher
Editorial Assistant: Kaitin Murphy
Vice President, Marketing: Jennifer Ann Baker
Marketing Director: Deborah Yarnell
Senior Marketing Manager: Erin Brennan
Associate Marketing Manager: Jillian Borden
Senior Production Director: Wendy A. Troeger
Production Manager: Mark Bernard
Content Project Manager: David Barnes
Production Technology Assistant: Emily Gross
Art Director: David Arsenault
Technology Project Manager: Joe Pliss
All cover images courtesy of iStockphoto.
Background Image: © CTR design LLC
Top left image: © Dean Turner
Center left image: © ryasick
Center right image: © Jakub Pavlinec
Bottom left image: © Lisa F. Young
Bottom right image: © michal kodym

ISBN-13: 978-1-111-63954-9

ISBN-10: 1-111-63954-X

Delmar
Executive Woods
5 Maxwell Drive
Clifton Park, NY 12065-2919
USA

Cengage Learning is a leading provider of customized learning solutions with office locations around the globe, including Singapore, the United Kingdom, Australia, Mexico, Brazil, and Japan. Locate your local office at **www.cengage.com/global**

Cengage Learning products are represented in Canada by Nelson Education, Ltd.

To learn more about Delmar, visit **www.cengage.com/delmar**

Purchase any of our products at your local bookstore or at our preferred online store **www.cengagebrain.com**

Printed in the United States of America
2 3 4 5 6 22 21 20 19 18

Contents

Preface

This laboratory manual is designed to accompany *Introduction to Low Voltage Systems*, 2nd Edition, by Amy DiPaola and Samuel DiPaola. The experiments and exercises are organized by chapter to accommodate a wide range of technical content from the text.

The lab manual uses a workbook format. Each assignment will include the following parts:

- *Required Reading*: reading assignments referenced to *Introduction to Low Voltage Systems*, 2nd Edition.
- *Introduction*: The introduction reinforces important information from the text as well as includes any additional information necessary to complete the assignment.
- *Pre-Job Questions*: The pre-job questions are to be completed prior to performing the lab procedure to determine the student's level of knowledge based on the required reading and introduction. The instructor should review pre-job questions before the students complete the remaining portions of the lab assignment.
- *Objectives*: A list of statements determining what the student should know after completing the lab assignment.
- *Required Equipment and Materials*: A list of required materials, tools, and test equipment needed to be able to complete the lab assignment.
- *Procedure*: A numerical procedure that logically steps the students through the lab assignment.
- *Final Summary*: The final summary focuses on analyzing the results of the lab procedure and reinforces the main points of the assignment.
- *Final Questions*: A list of questions to determine what the students learned from the assignment.

It is important to note that the instructor will be required to do some initial preparatory work on setting up some of the lab assignments; specifically in the chapters and assignments related to Security, Access-Control, and Fire Alarm Systems. For those labs, the lab assignments have been designed around the basic functions of a typical system. It is not the intent of this lab manual to teach to a specific manufacturer, system, or control panel. As a result, it will be the responsibility of the instructor to choose the type of system that will be used for the exercise. System programming, as well as any unique manufacturer specifics dealing with panel installation, will need to be determined and implemented by the instructor prior to attempting the lab procedures as written.

Lab Assignment Number and Title	Chapter Reference Guide from *Introduction to Low Voltage Systems*, 2nd Ed.
1 Learning to Use the *NEC*	Chapter 1–The National Electrical Code
2 Coaxial Cables and Connectors 3 Wire Resistance and Circular Mils	Chapter 2–Electrical Conductors and Cables
4 Grounding and Bonding	Chapter 3–Grounding and Bonding
5 dB, Voltage, and Power 6 Inverse Square Law	Chapter 4–Audio Physics
7 Installation of Ceiling-Mounted Class 2 or Class 3 Paging Speaker 8 Class 2 or Class 3 Audio Paging Systems	Chapter 5–Audio Signal-Processing, Amplification, and Reproduction Equipment
9 Networking Cables and Connectors 10 Crossover Cable	Chapter 6–Networking and Information Technology Equipment
11 Half-Wave Rectifiers and LEDs	Chapter 7–Power Supplies, Batteries, and Emergency Systems
12 Conduit Fill	Chapter 8–*Article 725* of the *National Electrical Code*®; Classification of Circuits
13 Fire Alarm Systems	Chapter 9–Fire Alarm Systems
14 Anerobic Fiber-Optic Connectors 15 Hot Melt® Fiber-Optic Connectors	Chapter 10–Fiber-Optic Cable and *National Electrical Code® Article 770*
16 Telephone Cables and Connectors 17 Telephone Local Loop Installation	Chapter 11–Telecommunications, and National Electrical Code, Article 800
18 Relays 19 Security Systems 20 Access Control Systems	Chapter 12–Security and Access Control System Basics
21 Installation of HD TV Antenna	Chapter 13–Wireless Communications and *National Electrical Code® Article 810*
22 Camera Systems	Chapter 14–Closed Circuit Television Camera Systems

Name ______________________

Date ______________________

Class ______________________

LAB 1

Learning to Use the *NEC*

REQUIRED READING

Introduction to Low Voltage Systems, 2nd edition, Chapter 1

INTRODUCTION

The following lab exercise is designed to help an individual new to the electrical industry become familiar with using *NFPA 70*, also known as, the *National Electrical Code* (*NEC*). To find and interpret *Code* references, it is important to understand the format, outline, and terminology used within the *NEC*. A series of *Code* references are listed in the Procedure section of this lab exercise, and you are asked to provide the section of the *Code* where the reference is located, as well as to complete the missing items within the statements. In addition, the exercise gives some practical examples of how to use the *NEC* index.

Being able to find and interpret electrical code is a critical and necessary skill for any individual in the electrical trade. The starting point for being able to maneuver through the *NEC* and find *Code* references is understanding how the book is formatted and logically laid out. Once you understand the structure of the *NEC*, finding and interpreting *Code* references becomes second nature.

Using the *NEC* Index

You can use the index at the back of the *NEC* to find a *Code* reference in a number of different ways. Let's say you want to find the *Code* reference for installing a charging station for storage batteries.

- You could look up *storage batteries* in the index, and then follow the list down to *garages*, which references *Section 511.10*.
- Or you could look up *battery charging equipment* in the index, which references *Section 511.10(A)*.

- You could also look up *garages*, to find that the first item in the list is *battery charging equipment*, which also references *Section 511.10(A)*.

To start any index search, you must first determine the main subject or item to be referenced. Having a secondary topic helps you break the search down into more detail. In the example above, the first search used *storage batteries* as a main reference, and *garages* as a second. In the second search, *battery charging equipment* was the main reference and a secondary reference was not needed. In the third search, *garages* was the main subject reference, and *battery charging equipment* was the secondary. Additional pathways may also exist for the example above; feel free to use an alternate manner, if you wish.

In the upcoming exercise, you are given a number of examples to look up on your own. By doing so, you will see that the index is a very powerful tool for finding unknown *Code* references.

PRE-JOB QUESTIONS

1. Turn to *NEC Section 90.3, Code Arrangement.*
   a. Which chapters apply generally to all electrical installations?
   b. Which chapters are supplemental?
   c. How does *Chapter 8, Communications Systems*, differ from the previous chapters?
2. Turn to *NEC Section 90.5, Mandatory Rules, Permissive Rules, and Explanatory Material.*
   a. Which terms in the *Code* are used for mandatory rules?
   b. Which terms in the *Code* are used for permissive rules, and how do they differ from mandatory rules?
   c. How is explanatory material referenced within the *NEC*?
   d. Are the informative Annexes enforceable requirements of the *NEC*?
3. Turn to *NEC Section 90.9, Units of Measurement.*
   a. What units of measurement are considered an International System of Units (SI)?
   b. When dual units of conversion are included in the *NEC*, which units are listed first, and which are placed in parentheses?

OBJECTIVES

- Become familiar with using the structure and format of the *NEC*.
- Be able to correctly interpret the differences between mandatory rules and permissive rules within the *NEC*.
- Locate unknown references within the *NEC* by using the index.

REQUIRED MATERIALS

2011 *National Electrical Code* book

PROCEDURE

Use the National Electrical Code (NEC) *to answer the following questions:*

1. There are ________ chapters within the *NEC*.
2. Each chapter is divided into ________.
3. Large articles of the *NEC* are divided into ________ designated by roman numerals.
4. Sections within the *NEC* are designated by ________.
5. How are changes from previous editions of the *NEC* identified in the current version?
6. Who has the authority to enforce the *Code*?
7. Who sponsors the *NEC*?
8. Formal interpretations of the *National Electrical Code* are permitted to be made by ________.
   a. the authority having jurisdiction (AHJ)
   b. National Electrical Manufacturers Association (*NEMA*)
   c. National Fire Protection Association (*NFPA*)
   d. Underwriters Laboratories, Inc. (*UL*)
9. The *NEC* is revised and updated every ________ years.

Find the answers to the next series of questions by using the NEC *index. Also include the section number where the* Code *reference was found.*

10. Overcurrent protection for Class 1 circuit conductors __________ AWG and larger _________ be provided in accordance with the conductor ampacity, without applying the ampacity adjustments and correction factors of *310.15* to the ampacity calculation.
    Section Ref. No. ___________
11. Overhead conductor clearance for communications, radio, and television coaxial cables within the scope of *Articles 800* through *820* shall be permitted at a height of not less than ________ feet above the swimming and wading pools, diving structures, and observation stands, towers, or platforms.
    Section Ref. No. ___________
12. A metal underground water pipe in direct contact with earth for _________ feet or more, and electrically continuous to the points of connection of the grounding electrode conductor and the bonding conductors or jumpers, if installed, shall be permitted as a grounding electrode.
    Section Ref. No. ___________
13. A separate grounding electrode for a CATV distribution system, in a one- or two-family dwelling, connected to a grounding electrode conductor longer than _________ feet in length, shall be as specified in *250.52(A)(5)*, *(A)(6)*, or *(A)(7)*.
    Section Ref. No. ___________

14. Where passing through a floor or wall, power-limited fire alarm (PLFA) conductors shall be installed in a _________ or _________ where passing through a floor or wall to a height of _________ feet above the floor, unless adequate protection can be afforded by building construction or unless a solid guard is provided.
    Section Ref. No. ___________
15. Find the definition for *conductive optical fiber cable*.
    Section Ref. No. ___________
16. Find the definition for *audio autotransformer*.
    Section Ref. No. ___________
17. Power supply cords for information technology equipment connected to a branch circuit shall not exceed ________ feet.
    Section Ref. No. ___________
18. Cables, raceways, or boxes installed in exposed or concealed locations under metal-corrugated sheet roof decking shall be installed and supported so there is not less than ________ inches measured from the lowest surface of the roof decking to the top of the cable, raceway, or box.
    Section Ref. No. ___________
19. Minimum burial depth for direct burial cables or conductors under streets, highways, roads, alleys, driveways, and parking lots shall be __________ inches.
    Section Ref. No. ___________
20. Where energized parts are exposed, the minimum clear workspace about equipment shall not be less than ________ feet high (measured vertically from the floor or platform) or not less than ________ feet wide (measuring parallel to equipment).
    Section Ref. No. ___________

FINAL SUMMARY

Being able to find *Code* references quickly and efficiently within the *NEC* is a critical skill to have when working in the electrical trade. A clear understanding of the structure of the book, as well as the format and terminology, is absolutely necessary in order to competently maneuver through the text. The NFPA revises the *NEC* every three years. Understanding how to use the *Code* book is imperative to being able to keep up with all the changes and possible additions that will be made in the coming years. Memorizing *Code* is simply not the answer; instead, every installer within the electrical trade should be able to independently look up the answers to any *Code*-related question and be able to site a reference if asked.

FINAL QUESTIONS

1. Were you able to prove for yourself that using the *NEC* index is a necessity when attempting to find unknown *Code* references?

2. Turn to *Article 800.44(B)* in the *NEC*. The general rule states that communications wires and cables shall have a vertical clearance of not less than ________ feet from all points of the roof above which they pass. Now look at *Exception No. 3*. Where the slope of the roof is not less than 4 inches, in 12 inches, a reduction in clearance to not less than ________ feet *shall be permitted*. Is *shall be permitted* a mandatory statement?
3. The *NEC* is written as a safety manual. Ultimately, who has the final say over any installation and how the *Code* is interpreted?
4. Turn to *Article 100* and locate the definition for *in sight from*. The definition states that a specified piece of equipment shall be visible from another, and not more than ________ feet distant from the other. If you had never read this definition, would you have had a clear understanding of what *in sight from* actually means? This example illustrates the importance of understanding definitions within the *NEC*.

Name ______________________

Date ______________________

Class ______________________

LAB 2

Coaxial Cables and Connectors

REQUIRED READING

Introduction to Low Voltage Systems, 2nd edition, Chapter 2

INTRODUCTION

The electrical installation requirement of coaxial cable is covered in *Article 820* of the *NEC*.

A diagram of coaxial cable is shown in Figure 2–1. The parts of a standard coaxial cable consist of an outer insulated jacket; a shield of braided metal; an aluminum-polyester foil; a flexible, internal foam insulation called a dielectric; and a central conductor. The central conductor is typically solid but in some cases can be stranded.

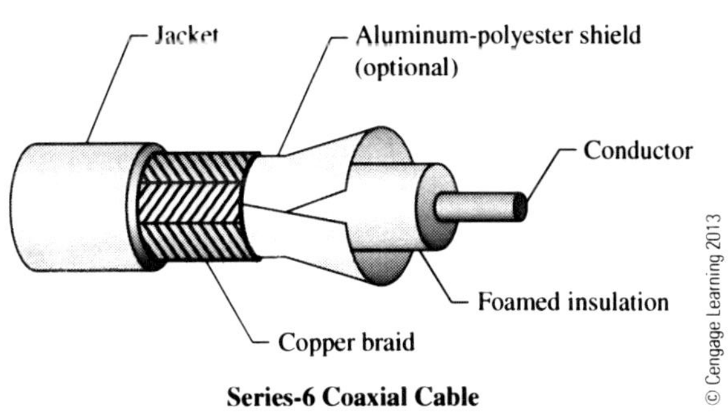

FIGURE 2–1 Parts of a coaxial cable.

TYPES OF COAXIAL CABLES

Depending on the type of system being installed, the choice of coaxial cable varies.

Video distribution uses a 75 ohm Series-6 or RG-6 coaxial cable, or type RG-59. Video Backbones uses a 75 ohm Series-11. Although the characteristic impedance of all of these cables is 75 ohms, the size of the central conductor varies: Series-11 is the largest, typically 14 AWG; Series-6, 18 AWG; and RG-59, 22 AWG.

Another common variety of coax is RG-8 and RG-58, which are used in radio antenna distribution. Both have a characteristic impedance of 50 ohms, with RG-8 having a larger central conductor, typically 10 AWG, and RG-58, 20 AWG.

For computer and instrumentation types of installations, a variety of coaxial cable called Series-62U or RG-62U also exists, which has a character impedance of 93 ohms.

Choosing the right type of cable depends on the type of system being installed and the impedance of the system. It is always critical to match the characteristic impedance of the cable to the impedance of the system. A mismatch of cable causes signal reflections to occur along the transmission line, resulting in increased system noise and distortion. Also, never interconnect different varieties of cable in series with each other, to do so also causes a mismatch and ultimately destroys the performance of the system.

TYPES OF CONNECTORS

This lab discusses the installation of standard BNC and F-type connectors. It is important to match the type of connector to the type of cable or the connector will not fit properly. When purchasing coaxial cable connectors, be sure to specify not only the type of cable but also the type of shield. A standard braided shield differs from a quad shield and therefore requires a different size connector. Connectors made for a standard braid Series-6 coax do not fit on a quad-shielded Series-6 coax, and likewise a quad shield connector is too large for a standard braid coaxial cable. In addition, you must also specify whether you are using plenum cable, because there is a difference between standard RG-59 connectors and RG-59 plenum connectors. Always know what type of cable you are installing, and be sure to correctly match the connector to the type and variety of cable.

PRE-JOB QUESTIONS

1. How does quad-shielded coaxial cable differ from a standard coaxial cable?
2. What is the difference between Series-6 and RG-58 coaxial cable?
3. What would occur if the characteristic impedance of the cable did not match the impedance of the system?

OBJECTIVES

- Recognize the characteristics and specifications of different types of coaxial cables.
- Demonstrate the installation techniques for BNC and F-type coaxial cable connectors.
- Demonstrate how to test the continuity of a coaxial cable with a digital multimeter.

REQUIRED MATERIALS

12 inches of Series-6 coaxial cable

12 inches of RG-59 coax

2 male, BNC-type connectors for Series-6 coaxial cable (2-piece or 3-piece)

2 male, F-type connectors for RG-59 coaxial cable (crimp style)

Coaxial cable stripping tool or wire stripper

Coaxial cable hex crimping tool for Series-6 and RG-59 cables

TEST EQUIPMENT

Digital multimeter (DMM)

PROCEDURE

Part I F-Type Connectors

1. Cut a 12-inch length of RG-59 coaxial cable and locate two F type connectors.
2. Locate the coaxial cable wire stripper. Images of tools are shown in the appendix. Depending on the diameter of the cable, you first have to adjust the depth of the cutting blades. When placing the cable into the stripper, be sure the end of the cable is flush with the edge of the stripper to ensure that you strip off no more than 1 inch of the end of the coax. The stripped cable should look like the cutaway shown in Figure 2–2. A regular wire stripper may also be used, but be sure not to nick the central conductor when cutting through the insulator, or signal degradation may occur.
3. Once the cable has been stripped, fold the braided shield back along the outer jacket, remove the exposed foil, and slide one of the F connectors onto the cable. If you are using a 2-piece F connector, be sure to slide

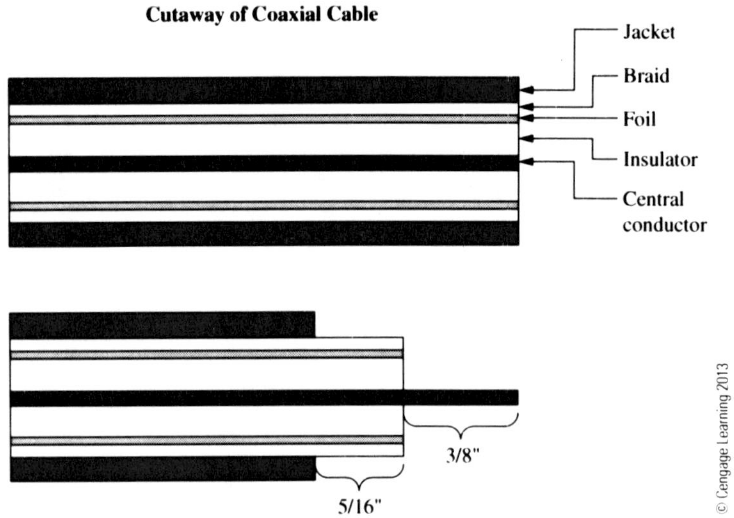

FIGURE 2–2 Cutaway of coaxial cable.

the crimp collar over the end of the stripped cable prior to attaching the connector. The F connector is shown in Figure 2–3.

4. Firmly push the end of the stripped cable into the connector until the white insulator is flush with the inside center of the connector and the central conductor is flush with the outside ring. Make sure that none of the shield wires are touching the central conductor when attaching the connector.
5. Using the hex crimping tool, crimp the neck of the F connector to securely fasten it to the RG-59 cable. If you are using a 2-piece connector,

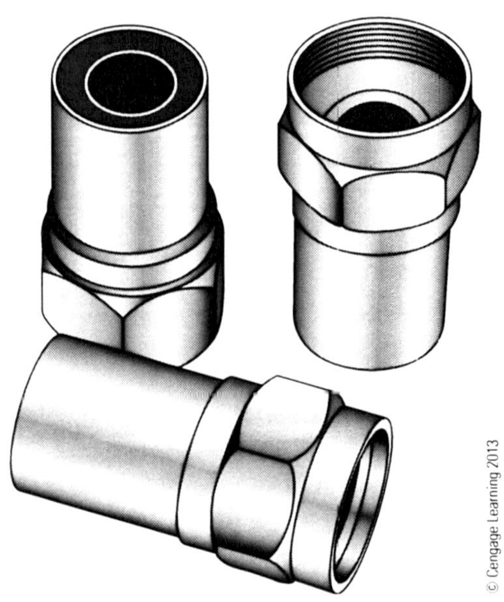

FIGURE 2–3 F connectors.

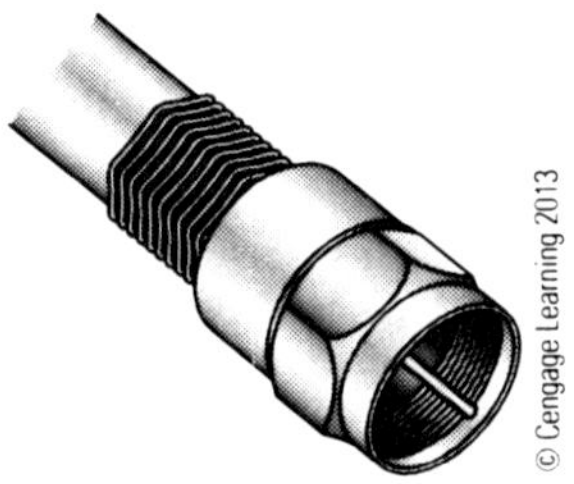

FIGURE 2–4 Crimped F connector.

slide the crimp collar up to the edge of the connector, and then crimp the connector into place with the hex crimping tool. Be sure the size of the crimp matches the type of cable. The crimped cable should look like the image in Figure 2–4.

6. Repeat steps 2–5 for the other side of the cable.
7. Test the cable for continuity with a digital multimeter (DMM). Measure the resistance between the central conductor and the outside of the F connector.

   ______________ ohms
8. If there is a short of zero ohms, then you need to try again. This indicates that the central connector is shorting the copper braid.

Part II BNC Connectors

9. Cut a 12-inch length of Series-6 coaxial cable, and locate two BNC connectors.
10. Use the coaxial wire stripper to strip one end of the cable, as described in Part I.
11. Fold back the shield and foil and trim the exposed insulator to 1/16 inch.
12. Now trim the exposed central conductor to 1/4 inch.
13. Locate the BNC connectors. You will either have a 2-piece or a 3-piece connector. The 3-piece contains a BNC connector, a central pin, and various sized outer crimp ferrules. An image of the 3-piece connector is shown in Figure 2–5. The 2-piece only contains a BNC connector along with various sized outer crimp ferrules, since the central pin is already installed inside the connector and does not need to be installed separately.
14. Manufacturers include various ferrule sizes to accommodate the different diameters of cable. Choose the correct size, based on the size of the cable you are using, and slide it onto the stripped end of the cable, narrow side first.

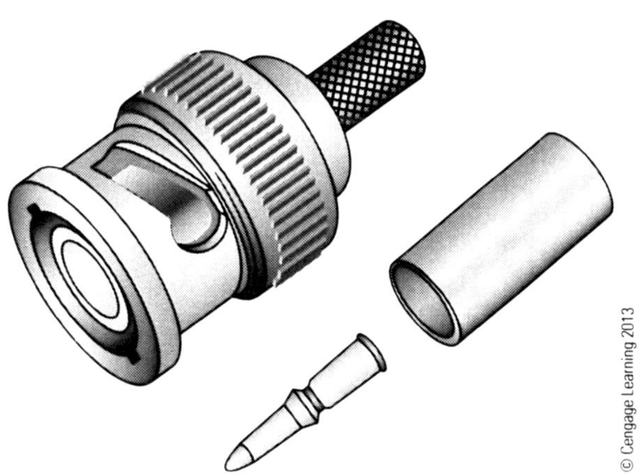

FIGURE 2–5 3-piece BNC connector.

© Cengage Learning 2013

FIGURE 2–6 Crimped BNC connector.

15. If you are using a 3-piece connector, you must first slide the center pin of the BNC connector onto the central conductor and crimp it into place. If you have a 2-piece connector, then skip this step.
16. Next, push the BNC connector onto the cable. If you are using a 3-piece connector, the end of the central conductor pin should push up nearly flush with the front of the white insulator of the BNC connector.
17. Slide the outer crimp ferrule over the back side of the BNC connector, covering the folded back portion of the braided shield, and crimp it into place with the hex crimping tool. Be sure the size of the crimp matches the size of the cable. The crimped connector should look like the image shown in Figure 2–6.
18. Repeat steps 13–17 for the other end of the cable.
19. Test the cable by measuring the resistance between the center pin and the outside of the BNC connector.

    _________ ohms
20. If there is a short of zero ohms, then you have to try again. This indicates that the central connector is shorting the copper braid.

FINAL SUMMARY

Coaxial cable is used for a variety of applications, including CATV, video distribution, cable television installation, radio antenna distribution, and even on some networked computer and instrumentation systems. The installation of BNC and F-type connectors is an important skill for any cable installer to master.

In addition, it is critical not only to know the specifications of and type of cable being used for an installation but also to correctly match and purchase the corresponding connecting hardware. A mismatch of cable and hardware can have undesirable consequences, resulting in standing wave reflections on the transmission line, which becomes a source of system noise, distortion, and signal loss.

FINAL QUESTIONS

1. Which was more difficult to terminate, the BNC or the F-type connector? Explain.
2. In which applications are coaxial cables commonly used?
3. If you do not know the specifications of a type of coaxial cable, how can you find out?
4. Why is it important to match the characteristic impedance of the cable to the impedance of the system?
5. Does the characteristic impedance of a coaxial cable change with length?

Name ______________________

Date ______________________

Class ______________________

LAB 3

Wire Resistance and Circular Mils

REQUIRED READING

Introduction to Low Voltage Systems, 2nd edition, Chapter 2

INTRODUCTION

$$R = K \times L/\text{C-mil}$$

where

R = Resistance

K = Ω per mil-foot constant, different for each type of conductor, and changes with temperature

L = Length

C-mil = Circular-mil area of a conductor

For voltage drop calculations, the formula can also be written as:

$$E = I \times K \times 2\ L/\text{C-mil}$$

L now represents length to the load. The "2" indicates that there are two wires, one traveling out to the load, and the other, back.

And because R = E/I, E and I can be substituted into the original equation.

Mil

1 mil is equal to 1/1000 of an inch, or 0.001 inches

Circular mil (C-mil)

$$(\text{Mil})^2 = \text{C-mil}$$

Resistivity constant

The "K" value for **copper**, at 20°C (68°F), is

10.4 Ω per mil-foot

The "K" value for **aluminum**, at 20°C (68°F), is

17 Ω per mil-foot

For most conductors, "K" values increase with temperature. For this lab, assume that the conductors are at 20°C (68°F), as they are in Table 3–1:

TABLE 3–1 The Table of American Wire Gauge (AWG) and Resistance

AWG Number	Diameter in Mils	Area in Circular Mils	Ω per 1000 ft of Copper at 20°C (68°F)	Ω per 1000 ft Aluminum at 20°C (68°F)
10	102	10,380	0.999	1.64
11	91	8,234	1.260	2.07
12	81	6,530	1.588	2.61
13	72	5,178	2.003	3.29
14	64	4,107	2.525	4.14
15	57	3,257	3.184	5.22
16	51	2,583	4.016	6.59
17	45.3	2,048	5.06	8.31
18	40.3	1,624	6.39	10.5
19	35.9	1,288	8.05	13.2
20	32	1,022	10.15	16.7

A decrease in 3 AWG numbers doubles the cross-sectional area of the conductor and cuts the DC resistance in half.

PRE-JOB QUESTIONS

1. In wire measurements, what is a mil?

   What is a circular mil?

   What is a mil-foot?
2. Calculate circular mil cross-sectional area of a wire 0.012 inches in diameter.
3. Find the diameter of a wire that has an 81 C-mil cross-sectional area.
4. Find the resistance of a #14 AWG aluminum wire, 100 feet in length, using

$$R = K \times L/C\text{-mil}$$

5. A power line is to use 15,000 feet of wire. The total resistance should not exceed 5 ohms. What diameter copper wire meets these requirements?

OBJECTIVES

- Calculate and determine the unknown length of a conductor based on its measured resistance value.
- Calculate and determine the expected voltage drop on a measured length of cable with varying loads.

REQUIRED MATERIALS

A spool of 2-conductor, 18 AWG copper cable, having an unknown length, estimated between 500 and 1000 feet (you must have access to both ends of the spool)

Two 47-ohm resistors

Two 100-ohm resistors

PAD—234 analog digital trainer

TEST EQUIPMENT

Digital multimeter (DMM)

PROCEDURE

1. Obtain the spool of wire from the instructor. Be sure that you have access to both ends of the cable on the spool.
2. Strip the wires on each end of the spool using a wire stripper.
3. Connect each end of the spool with jumper wires, to points A, B, C, and D of Figure 3–1. *Be careful that the ends of the wire spool do not accidentally short together, or the PAD trainer may be damaged.* For now, **keep the power off**.

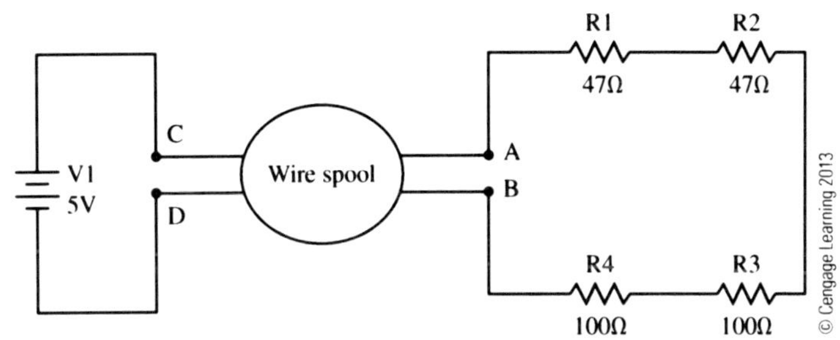

FIGURE 3–1

Steps 4–11 are repeated two times. ✓ Check off each step as you perform it to keep track of where you are in the procedure.

1✓ 2✓

4. Disconnect the wires from the 5-volt supply and measure the resistance of the entire circuit, using a DMM at points C and D. Enter the value in Table 3–2 as R_{CD} for Figure 3–1.

   1__ 2__

5. Reconnect the wires to the 5-volt supply.

   1__ 2__

6. Next, disconnect the spool from points A and B.

   1__ 2__

7. Using a DMM, measure the total load resistance at points A and B. Record the value as R_{AB} in Table 3–2 for Figure 3–1.

   1__ 2__

8. Reconnect the spool to A and B, and turn on the power to the PAD trainer.

   1__ 2__

9. Using a DMM, measure V_{AB}. Record the value in Table 3–2 for Figure 3–1.

   1__ 2__

10. Measure V_{CD}. Record the value in Table 3–2 for Figure 3–1.

    1__ 2__

11. Shut off the power.

    1__ 2__

TABLE 3–2

	Figure 3–1	Figure 3–2
5-volt supply V_{CD}		
Load voltage V_{AB}		
Cable voltage drop $V_{CD} - V_{AB}$		
R_{CD}		
R_{AB}		
Total current $I_T = V_{CD}/R_{CD}$		

12. Build circuit 2, as shown below in Figure 3–2.

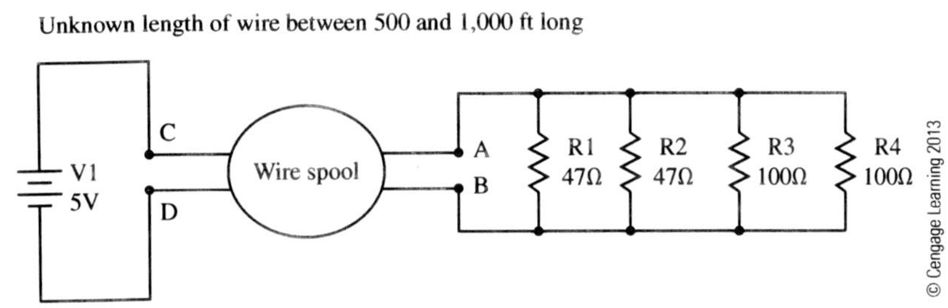

FIGURE 3–2

13. Repeat steps 4–11 for Figure 3–2.

CALCULATIONS

14. Using the data in Table 3–2, subtract V_{AB} from V_{CD} for each circuit. Record the results as cable voltage drop in Table 3–2 for Figures 3–1 and 3–2.

$$V_{CD} - V_{AB} = \text{Cable Voltage Drop}$$

15. Using the data in Table 3–2, and the following formula,

$$V_{CD}/R_{CD} = I_T, \text{ Total Current}$$

calculate the total current flow for each circuit. Record the results as total current in Table 3–2 for Figures 3–1 and 3–2.

16. The next step is to calculate the length of the cable by using the data from Table 3–2. Starting with the formula:

$$R = K \times L/\text{C-mil}$$

Solving for L:

$$L = R \times \text{C-mil}/K$$

Using the data in Table 3–2, calculate the cable resistance, ***R***, for the above formula, by subtracting R_{AB} from R_{CD}.

$$R = R_{CD} - R_{AB}$$

Enter results in Table 3–3.

TABLE 3–3

	Figure 3–1
R_{CD} (taken from Table 3–2)	
R_{AB} (taken from Table 3–2)	
Measured cable resistance, R	

Obtain the value of **C-mil** from Table 3–1: Table of American Wire Gauge (AWG) and Resistance.
C-mil = __________

For copper cable, assuming that the lab environment and cable temperature are approximately 20°C (68°F), use a K value of 10.4 ohms per mil foot.

Now that we have determined the value of R, C-mil, and K, solve for L:

L = _____ feet (length of connecting loop)

Now divide by 2 to obtain the **length to the load**, because the above calculation represents the total length of the connecting loop, out to the load and back.

Length to the load __________ feet

17. Calculate the cable voltage drop:

$$E = I \times K \times 2\ L/\text{C-mil}$$

Be sure to use the same K and C-mil values when calculating the E for each circuit.

Use the length to the load, determined from step 16, and the total current, I, as taken from Table 3–2 for each figure. Enter the values for each figure in Table 3–4.

TABLE 3–4

	Figure 3–1	Figure 3–2
Total current, I_T (taken from Table 3–2)		
Calculated cable voltage drop: $E = I \times K \times 2\ L/\text{C-mil}$		

FINAL SUMMARY

When designing an installation, it is sometimes necessary to be able to calculate the expected voltage drop on a length of cable. The result may ultimately determine the maximum length of cable to the load, the maximum number of parallel loads on a cable, and the cross-sectional area or AWG of the conductor to be used.

A measurable voltage drop on a length of cable means that there is a loss of voltage at the load.

$$E_{Source} = E_{Cable} + E_{Load}$$

or

$$E_{Load} = E_{Source} - E_{Cable}$$

Circuits are purposely designed to minimize the amount of voltage drop over the length of cable in order to maximize the amount of power being delivered to the load. To do so increases the efficiency of the circuit and also helps to reduce excessive demands on the power supply.

FINAL QUESTIONS

1. Work out the calculated voltage drops in Table 3–4, and compare them to the measured voltage drops in Table 3–2.
2. From the data in Table 3–2, what happens to the cable voltage drop as the total current increases?
3. From the data in Table 3–2, what happens to the load voltage, V_{AB}, as circuit current increases?
4. From the data in Table 3–2, as cable voltage drop increases, what happens to the load voltage?
5. The sum of the cable voltage drop and the load voltage must equal what circuit variable?
6. If the length of the connecting cable were to double, what effect would it have on the circuit?
7. If the AWG of the connecting cable were to increase, what effect would it have on the circuit?
8. To increase the cable length and not lose voltage at the load, what would be an obvious solution?

Name ______________________

Date ______________________

Class ______________________

LAB 4

Grounding and Bonding

REQUIRED READING

Introduction to Low Voltage Systems, 2nd edition, Chapter 3

INTRODUCTION

The purpose of this lab exercise is to introduce you to the assembly methods and installation of an intersystem bonding termination busbar (IBTB), a telecommunications main grounding busbar (TMGB), or a telecommunications grounding busbar (TGB). After completing the assembly of the busbars, you cut various lengths of 6 AWG copper conductor and terminate them onto single-hole or 2-hole compression connectors. The terminated conductors are then connected to the appropriate busbars, a metal equipment enclosure, and a short piece of ⅝-inch ground rod via a ⅝-inch ground clamp, to simulate making connections for bonding conductors, equipment grounding conductors, and grounding electrode conductors.

To complete this lab, you will need a lab booth or wall space that has a piece of 4 × 8 × ¾-inch plywood vertically mounted. The lab exercise then simulates the grounding of a communication system within a two-floor structure.

The Intersystem Bonding Termination Busbar

Busbars are made of copper and come in a variety of widths and lengths. The telecommunications main grounding busbar (TMGB) represents the largest, typically ¼ inch thick, 4 inches wide, and 8 to 12 inches long. The TMGB is the main busbar mounted on the wall of the equipment room in the lower level of the building, typically where the service providers bring the outside cabling into the structure.

The telecommunications grounding busbar (TGB) is usually smaller in size, typically ¼ inch thick, 2 inches wide, and 6 to 12 inches long. The TGBs are mounted in telephone or equipment room closets on the floors above the main level. In addition, the TMGB and TGB shall be insulated from their supports.

A minimum separation of 2 inches from the wall is recommended. Mounting kits purchased with busbars typically provide the insulators as well as the necessary 2-inch separation. Figures 4–1 and 4–2 illustrate a typical TMGB and TGB.

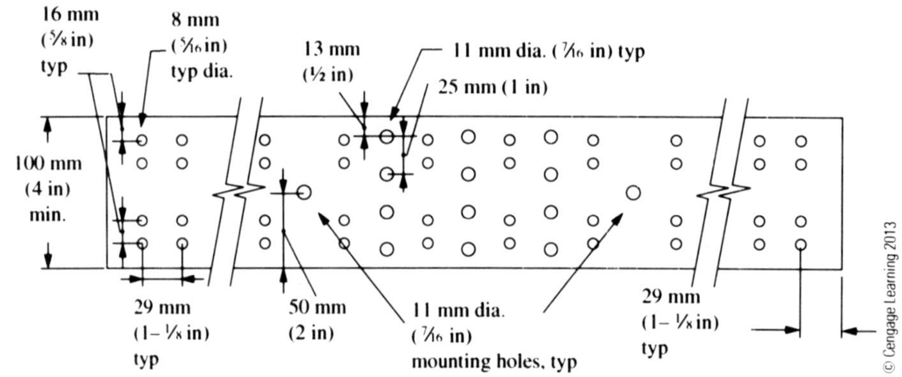

FIGURE 4–1 Typical telecommunications main grounding busbar (TMGB).

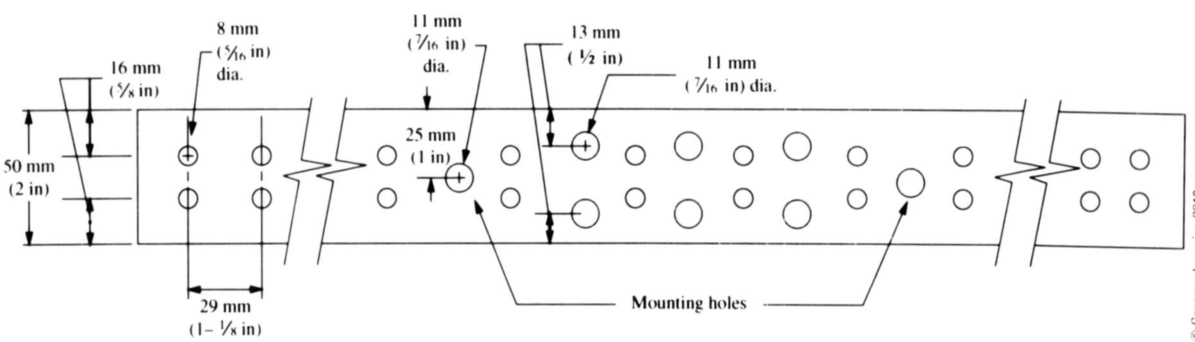

FIGURE 4–2 Typical telecommunications grounding busbar (TGB).

Once mounted, the TMGB is then bonded to the electrical service entrance of the building, as indicated in Figure 4–3.

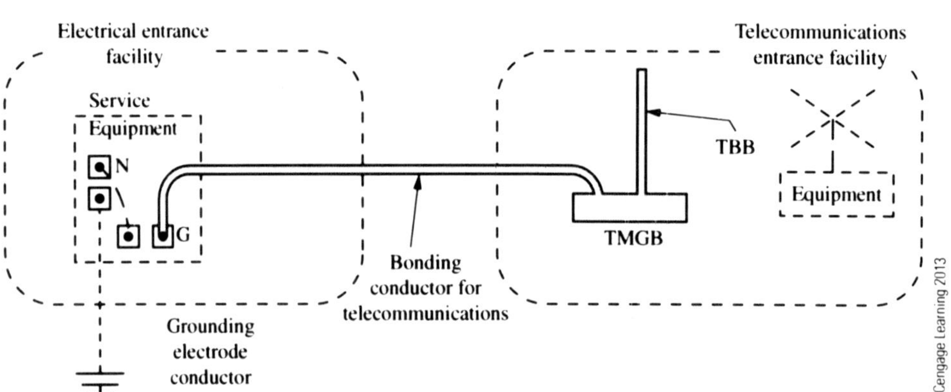

FIGURE 4–3 Bonding to the service equipment (power) ground.

TBB represents the telecommunications bonding backbone, which interconnects to all of the TGBs within the building.

Figure 4–4 illustrates what a completed grounding system would look like for a large commercial building.

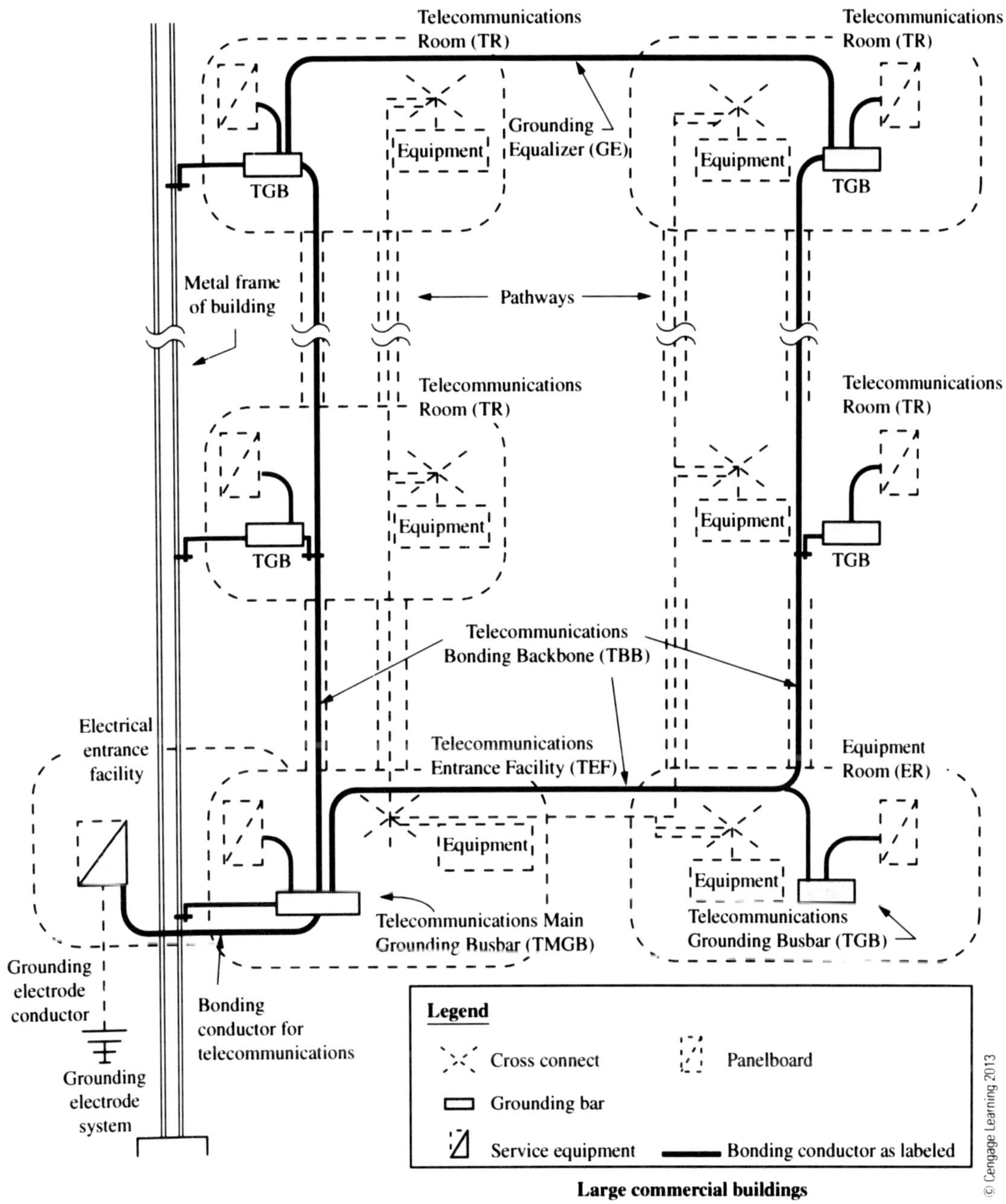

FIGURE 4–4

The illustration shows a three-floor building. Notice how all of the TGBs are interconnected by the TBB. Notice too that the TGBs are also interconnected to building steel as well as to all the panelboards within each telecommunications closet. Also, the bonding conductor on the top floor that links between the two sides of the building is called a grounding equalizer (GE). Once completed, the telecommunications grounding bus interconnects all electrical systems within the building and all exposed metal, as well as building steel, into a single, interconnected ground plane.

The size of the bonding conductors is determined by the length of the cable run, as indicated by Table 4–1. For high-frequency installations, the size of the bonding conductors should be increased to accommodate for losses due to skin effect.

TABLE 4-1 Sizing of the TBB

TBB Length Linear m (ft)	TBB Size (AWG)
less than 4 (13)	6
4–6 (14–20)	4
6–8 (21–6)	3
8–10 (27–33)	2
10–13 (34–41)	1
13–16 (42–52)	1/0
16–20 (53–66)	2/0
greater than 20 (66)	3/0

Telecommunications equipment is then bonded to the busbars as needed. The number of connections needed for a given bar determines the size and number of holes drilled into the copper plate.

For this lab exercise, we use a 6 AWG copper conductor; solid or stranded is fine, and bare or green insulated.

PRE-JOB QUESTIONS

When answering the following questions, please reference the NEC.

1. The primary protector for a communications cable entering a building from outside shall be located _____________ to the point of entrance.
2. Where a separate communications grounding electrode is installed, it shall be bonded back to the main power grounding electrode system with a minimum _____________ AWG copper conductor.
3. What types of connections can be used when connecting a bonding conductor or grounding electrode conductor to a grounding electrode?
4. What types of connections cannot be used when connecting a bonding conductor or equipment grounding conductor to equipment?

OBJECTIVES

- Demonstrate the installation of an intersystem bonding termination busbar (IBTB), telecommunications main grounding busbar (TMGB), or telecommunications grounding busbar (TGB).

- Demonstrate the connection of a bonding conductor or grounding electrode conductor to an IBTB, TMGB, or TGB.
- Demonstrate the connection of a grounding electrode conductor to a rod-type grounding electrode.
- Demonstrate the connection of an equipment grounding conductor to a communications equipment rack or metal equipment enclosure.

REQUIRED MATERIALS

Safety glasses

Copper busbar for TMGB, having minimum dimensions of 6 mm (0.25 in.) thick × 100 mm (4 in.) wide × 152 mm (6 in.) long

Copper busbar for TGB, having minimum dimensions of 6 mm (0.25 in.) thick × 50 mm (2 in.) wide × 152 mm (6 in.) long

2 busbar mounting kits, each including 2 busbar insulators, 2 busbar mounting brackets, and supporting hardware

1 tinned copper, single-hole compression lug for AWG 6 solid or stranded copper conductors

4 tinned copper, 2-hole compression lugs for AWG 6 solid or stranded copper conductors

Compression tool for compression lugs

⅝-inch-diameter copper ground rod, 3 feet in length

Two 2½-inch D-rings

Four ¾-inch-long mounting screws for the D-rings

⅝-inch ground rod clamp

8 feet of AWG 6 copper conductor, solid or stranded, bare or green insulated

1 metal equipment enclosure (15 in. H × 12 in. W × 4 in. D, or something similar in size)

Eight 5/16-inch-diameter lag bolts, ¾ inch long

Nine 5/16-inch-diameter bolts, ¾ inch long; lock washers and nuts

PROCEDURE

Part I Installation of Intersystem Bonding Terminal Bars

1. Locate the two mounting holes on one of the copper busbars.
2. Attach the two insulators to the copper busbar, using the hardware included in the mounting kits.
3. Attach a mounting bracket to the opposite side of each insulator, using the hardware included in the mounting kit. Be sure that the mounting brackets are positioned perpendicular to the busbar prior to tightening.
4. Assemble the second busbar in like manner.
5. Locate four, ¾-inch long-lag bolts.

6. Position the assembled TMGB busbar (the 100-mm-[4 in.] wide bar) centered within the lab booth space or on the plywood, horizontally, 4 feet from the floor.
7. Attach the TMGB busbar to the plywood using the four, ¾-inch-long lag bolts.
8. Locate four more, ¾-inch-long lag bolts.
9. Position the assembled TGB busbar (the 50-mm-[2 in.] wide bar) centered within the lab booth space or on the plywood, horizontally, 2 feet above the TMGB.
10. Attach the TGB busbar to the plywood using four of the ¾-inch-long lag bolts.

Part II Connection of Bonding Conductor between Intersystem Bonding Terminal Bars

11. Cut a piece of 6 AWG copper conductor to a length of 2½ feet, to be used as a bonding conductor between the TMGB and the TGB.
12. Locate two 2-hole compression connectors.
13. Using the compression tool, attach a 2-hole compression connector to each end of the bonding conductor.
14. Now attach one end of the bonding conductor to the TMGB using two 5/16-inch-diameter bolts, lock washers, and nuts.
15. Extend the other end of the bonding conductor up to the TGB, and attach it using two more 5/16-inch-diameter bolts, lock washers, and nuts.

Part III Connection of Equipment Grounding Conductor to Communications Equipment Rack or Metal Equipment Enclosure

16. Mount a metal equipment enclosure 1 foot to the right of the TGB.
17. Cut a piece of 6 AWG copper conductor to a length of 2½ feet, to be used as an equipment grounding conductor.
18. Locate a single-hole compression connector and attach it to one end of the equipment grounding conductor using the compression tool.
19. Locate a 2-hole compression connector and attach it to the other end of the equipment grounding conductor, using the compression tool.
20. Drill an 8-mm (5/16-in.) hole into the lower left side of the metal equipment enclosure.
21. If the equipment enclosure is painted, be sure to sand off the paint from around the drilled hole, on the inside and outside of the enclosure, to ensure that a solid connection is made when connecting the equipment grounding conductor.
22. Attach the single-hole connector end of the equipment grounding conductor to the metal equipment enclosure, using a 5/16-inch-diameter bolt, lock washer, and nut.
23. Attach the 2-hole connector end of the equipment grounding conductor to the TGB using two 5/16-inch-diameter bolts, lock washers, and nuts.

Part IV Connection of Grounding Electrode Conductor to the Ground Rod

24. Mount two 2½-inch D-rings, 2 feet apart, vertically, 1 foot to the left of the TMGB. The D-rings are used to vertically support the segment of copper ground rod.
25. Insert a 3-foot section of copper ground rod vertically between the D-rings.
26. Cut a 3-foot section of 6 AWG copper conductor to be used as a grounding electrode conductor.
27. Locate a 2-hole compression connector.
28. Using the compression tool, attach one end of the grounding electrode conductor to the 2-hole compression connector.
29. Now attach the 2-hole connector to the TMGB using two 5⁄16-inch-diameter bolts, lock washers, and nuts.
30. Attach the other end of the grounding electrode conductor to the upper end of the ⅝-inch ground rod using the ⅝-inch ground clamp.
31. Have the instructor check all of your connections for parts I–IV of this lab.

FINAL SUMMARY

After completion of this assignment the assembly, within the lab booth or on the plywood should look like the illustration in Figure 4–5.

TIA/EIA J-STD-607-A covers the commercial building grounding and bonding requirements for telecommunications. Please reference this standard for additional information on telecommunications system grounding and bonding.

Also, the TMGB or TBG should be placed as close as practicable to the panel board within the room to which it is installed.

The connections of the bonding conductor, equipment grounding conductor, grounding electrode conductor, and the TBB to the TMGB or TGB shall utilize exothermic welding, listed compression connectors (2-hole lugs are preferred), or other irreversible compression-type connections. Twisting-type lugs shall not be used.

Splices to the TBB shall be avoided. Any connections to the TBB shall be made by exothermic welding or through irreversible compression-type connectors, such as listed C- or H-type connectors.

All metallic raceways for telecommunications equipment located within the same room or space as the TMBG or TGB shall be bonded to the TMBG or TGB as well.

An additional item of note, where a bonding conductor, equipment grounding conductor or grounding electrode conductor is run within a metal raceway, it must also be bonded to the raceway through which it travels, at both ends.

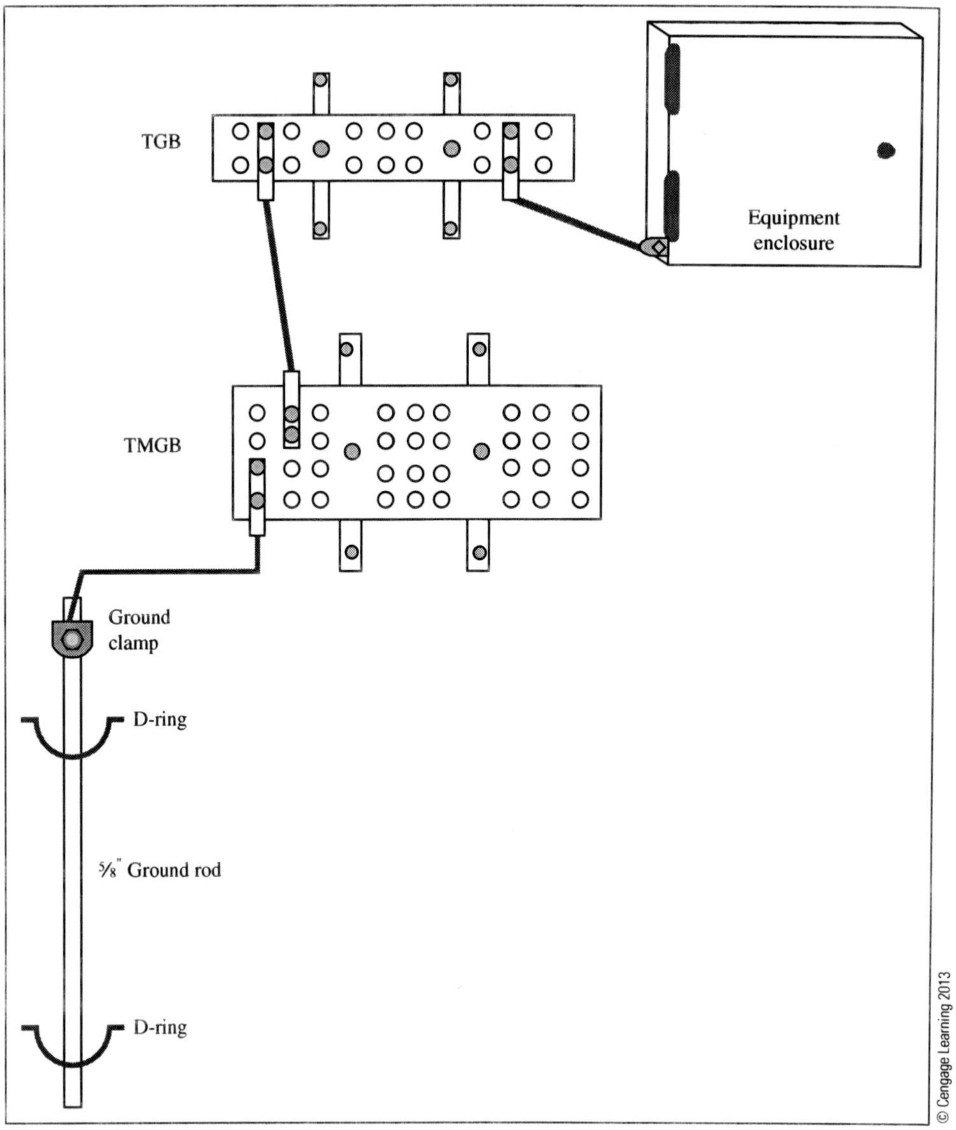

FIGURE 4–5

FINAL QUESTIONS

1. When installing the TMGB or TGB, a minimum separation of ________ inches from the wall is recommended.
2. When providing ground to a high-frequency system, what needs to be done to the size of the bonding conductor, and why?
3. Where is a TMGB located, and what is connected to it?
4. Where are TGBs typically located, and how do they differ from TMGBs?

Name ______________________

Date ______________________

Class ______________________

LAB 5 dB, Voltage, and Power

REQUIRED READING

Introduction to Low Voltage Systems, 2nd edition, Chapter 4

INTRODUCTION

The design of any audio system ultimately comes down to determining the required decibel levels needed for a specific sized space. Whether it be for a small home theater, a large commercial auditorium, or an outdoor stadium, a clear understanding of the relationships between power, voltage, and decibel is essential to obtaining satisfactory results.

In this lab experiment, you connect a simple audio system and measure how changes in voltage and power affect the decibel output of the amplifier.

During the lab procedure, you are also asked several times to swap out an 8-ohm speaker for an 8-ohm load resistor, and vice versa. This is because the inductive properties of the 8-ohm speaker generate errors in the voltage readings if the speaker remains connected to the output of the amplifier during those measurements. Accuracy in this experiment can therefore be guaranteed only by measuring voltage across a fixed 8-ohm load resistor instead of directly across an active speaker.

The decibel measurements are taken when the speaker is connected, and the voltage measurements are taken only when the resistor is connected.

Logarithm basics:

$$100 = 10^2$$

$$\log 100 = 2$$

Therefore, a log is an exponent of a base 10 number.
Another example:

Because $\log 700 = 2.845$, then $10^{2.845} = 700$.

PRE-JOB QUESTIONS

1. Decibel is a linear scale. True or false?
2. From a sound and hearing point of view, what does 0 dB represent?
3. The threshold of pain is equal to _______ dB.
4. A doubling of amplifier power increases the output by _______ dB.
5. A doubling of output voltage increases the output by _______ dB.

OBJECTIVES

- Measure the voltage and decibel outputs of an audio system.
- Determine how a doubling of voltage or power effects decibel levels.

REQUIRED MATERIALS

8-ohm speaker

3-foot section of 18 AWG speaker wire

8-ohm, 25-watt resistor

Two alligator clips

Connecting cable: RCA to Banana or BNC (depending on what type of output connector is on the audio generator)

TEST EQUIPMENT

Radio Shack digital sound level meter

Digital multimeter (DMM)

10-watt audio amplifier

Audio generator

PROCEDURE

1. Adjust the audio generator to 1000 Hz.
2. Set the output voltage of the audio generator to 100 mV. Verify the level with a DMM connected to the output of the generator (be sure that the meter is set to measure ac volts).
3. Connect the audio generator to an input of the audio amplifier. You need either an RCA to Banana or RCA to BNC cable, depending on what type of output is installed on the audio generator.
4. Using alligator clips, connect an 8-ohm load resistor to the appropriate 8-ohm output of the audio amplifier.
5. Connect the DMM across the 8-ohm load of the amplifier.

6. Turn on the amplifier and adjust the amplifier volume control until the output voltage measures 2.8 volts.
7. Using the formula V^2/R, calculate the power at the output of the amplifier.
   ____________ watts
8. Turn off the amplifier, but do not change the level of the volume control.
9. Disconnect the resistor load and attach the 8-ohm speaker to the output of the amplifier.
10. Turn on the digital sound level meter, and adjust it for a range of 80 dB, using weighting curve A.
11. Place the sound level meter 3 feet or 1 meter from the speaker. Turn on the amplifier and measure the sound level.
    ____________ dB (with a 1-watt output)
12. Slowly increase the volume control of the amplifier until the sound level meter recognizes an increase of 3 dB.
    ____________ dB
13. Turn off the amplifier, being sure not to change the level of the volume control.
14. Disconnect the speaker, and reattach the 8-ohm load resistor.
15. Turn on the amplifier and measure the voltage of the 8-ohm load, using the DMM.
    ____________ volts
16. Using the formula V^2/R, calculate the power at the output for the new voltage reading.
    ____________ watts
17. Turn off the amplifier, being sure not to change the level of the volume control.
18. Disconnect the load resistor and reattach the speaker.
19. Turn on the amplifier.
20. Place the sound level meter 3 feet or 1 meter from the speaker. It should indicate the previous level of dB.
21. Adjust the volume control of the amplifier until the output increases an additional 3 dB.
    ____________ dB
22. Turn off the amplifier, being sure not to change the level of the volume control.
23. Disconnect the speaker and reattach the 8-ohm load resistor.
24. Turn on the amplifier and measure the voltage of the 8-ohm load, using the DMM.
    ____________ volts
25. Using the formula V^2/R, calculate the power at the output for the new voltage reading.
    ____________ watts
26. Turn off all equipment.

FINAL SUMMARY

The formula for calculating the decibel equivalent of a power measurement is

$$dB = 10 \log P/P_{Ref}$$

with P_{Ref} representing the reference power value.

The formula for calculating the decibel equivalent of a voltage measurement is

$$dB = 20 \log E/E_{Ref}$$

with E_{Ref} representing the reference voltage value.

A doubling of power increases the output by +3 dB.

A doubling of voltage increases the output by +6 dB.

FINAL QUESTIONS

1. Calculate the dB increase of the amplifier from the initial 1-watt output to the value obtained in step 16.
2. Calculate the dB increase of the amplifier from the initial 2.8 volts to the value measured in step 24.
3. Do your calculated dB values come close to the expected values for when power and voltage are doubled?

Name ______________________

Date ______________________

Class ______________________

LAB 6

Inverse Square Law

REQUIRED READING

Introduction to Low Voltage Systems, 2nd edition, Chapter 4

INTRODUCTION

The inverse square law is used to determine the expected decibel level of an audio output at various distances from a source. The law states that signal intensity shall be inversely affected by the square of the changing distance. For example, if the distance were to double, then the signal intensity would be reduced by a factor of four ($2^2 = 4$). Logarithmically, a factor of four represents a change of 6 dB. Therefore, if an audio signal measures 55 dB at a distance of 5 feet (1.5 m) from a source, the same signal measures 49 dB at a distance of 10 feet (3 m) from the source.

In this experiment, to prove the inverse square law, you connect a simple audio system and measure the sound intensity at distances of 1, 2, 4, and 8 feet from the speaker.

PRE-JOB QUESTIONS

1. Besides audio, what other types of signals can be applied to the inverse square law?
2. What physical reason is there for why the signal intensity decreases by the square of the changing distance?

OBJECTIVES

- Measure the decibel output of an audio system at various distances from a speaker to prove the inverse square law.

REQUIRED MATERIALS

8-ohm speaker

3-foot section of 18 AWG speaker wire

8-ohm, 25-watt resistor

Two alligator clips

Connecting cable: RCA to Banana or BNC (depending on what type of output connector is on the audio generator)

TEST EQUIPMENT

Radio Shack digital sound level meter

Digital multimeter (DMM)

10-watt audio amplifier

Audio generator

PROCEDURE

1. Adjust the audio generator to 1000 Hz.
2. Set the output voltage of the audio generator to 100 mV. Verify the level with a DMM connected to the output of the generator (be sure that the meter is set to measure ac volts).
3. Connect the audio generator to an input of the audio amplifier. You need either an RCA to Banana or RCA to BNC cable, depending on what type of output is installed on the audio generator.
4. Using alligator clips, connect an 8-ohm load resistor to the appropriate 8-ohm output of the audio amplifier.
5. Connect the DMM across the 8-ohm load of the amplifier.
6. Turn on the amplifier and adjust the amplifier volume control until the output voltage measures 2.8 volts.
7. Turn off the amplifier, but do not change the level of the volume control.
8. Disconnect the resistor load and attach the 8-ohm speaker to the output of the amplifier.
9. Turn on the digital sound level meter, and adjust it for a range of 80 dB, using weighting curve A.
10. Place the sound level meter 1 foot from the speaker. Turn on the amplifier and measure the sound level.

    ______________ dB
11. Move the sound level meter to a distance of 2 feet from the speaker and measure the sound level.

    ______________ dB

12. Move the sound level meter a distance of 4 feet from the speaker and measure the sound level.

 _____________ dB

13. Move the sound level meter a distance of 8 feet from the speaker and measure the sound level.

 _____________ dB

14. Turn off all equipment.

FINAL SUMMARY

The inverse square law states that the measured intensity of an audio signal decreases by 6 dB every time the distance from the speaker is doubled.

To mathematically prove the inverse square law, let's calculate the expected change in signal intensity between steps 10 and 13 of the above procedure. Initially, the signal meter was placed 1 foot from the speaker. In step 13, the distance was increased to 8 feet, or 8 times the initial distance.

$$8^2 = 64$$

Therefore, the expected decibel level at step 13 should have decreased by

$$10 \log 64 = 18 \text{ dB}$$

Or we could have calculated it as

$$20 \log 8 = 18 \text{ dB}$$

FINAL QUESTIONS

1. Calculate the predicted dB change in the above procedure at a distance of 6½ feet from the speaker.
2. What signal level does the sound level meter indicate at a distance of 6½ feet?
3. Do the measured dB values come close to the expected values?

Name ______________________

Date ______________________

Class ______________________

LAB 7

Installation of Ceiling-Mounted Class 2 or Class 3 Paging Speaker

REQUIRED READING

Introduction to Low Voltage Systems, 2nd edition, Chapter 5

INTRODUCTION

The majority of class 2 or class 3 audio paging speakers are installed in a suspended ceiling. This lab exercise outlines the procedure for installing 8-inch audio paging speakers into a removable 24-inch ceiling tile. Tile sizes are typically 24 inches × 24 inches or 24 inches × 48 inches.

The installation procedure consists of 4 basic steps:

- Preparing the ceiling tile
- Preparing the back box
- Determining the connection to the speaker
- Putting it all together

PRE-JOB QUESTIONS

1. Does the *NEC* allow cables and wires to lie on top of suspended ceiling tiles?
2. What size ceiling tile will you use for this lab?

OBJECTIVES

- Identify the components of a ceiling-mounted speaker assembly.
- Demonstrate the preparation of a ceiling tile for installation of a speaker assembly.
- Demonstrate the final assembly and installation of a paging speaker within a suspended ceiling tile.

REQUIRED MATERIALS

70- or 25-volt, 8-inch ceiling speaker, with 13-inch grill cover, and 4 mounting screws

Atlas 81-8R 24-inch ceiling tile bridge (also known as a T-bar) with round hole

Back box enclosure for 8-inch speaker—the mounting diameter should measure 12 inches

24- × 48-inch or 24- × 24-inch ceiling tile

Drywall or keyhole saw

Wire strippers

1-inch bushing or standard Romex connector

PROCEDURE

Preparing the Ceiling Tile

1. Place the ceiling tile front side up, over a trash can.
2. The tile bridge is used as a template for marking the location of the speaker hole in the ceiling tile. Place the tile bridge over the ceiling tile in the desired location, squaring it up across the 24-inch width.
3. Trace the location of the center hole of the tile bridge onto the ceiling tile.
4. Next, identify the four mounting holes, located around the center hole of the bridge. These holes match up with the four mounting holes of the speaker-grill assembly and the back box.
5. Using a small-diameter screwdriver or awl, punch the location of the four mounting holes to mark their position on the ceiling tile.
6. Remove the tile bridge from the ceiling tile.
7. Using a drywall or keyhole saw, cut out the hole for the speaker, allowing the pieces of the ceiling tile to fall into the trash can.

Preparing the Back Box

8. Locate the back box.
9. There should be four 1-inch knockouts spaced around the sides of the back box. Determine which knockout will be used for feeding the speaker wire into the enclosure.

10. Punch out the desired knockout and insert a 1-inch bushing or standard Romex connector. The bushing or Romex connector is used to protect the cable from possible abrasion during the installation process.

Determining the Connection to the Speaker

11. Locate the transformer on the back of the speaker-grill assembly.
12. For this lab, you can use either a 70-volt or 25-volt speaker transformer. Which one are you using?
    __________ volt
13. The wires attached to the transformer are marked as follows: Com, 5W, 2W, 1W, ½ W, and ¼ W.

    Determine which tap connections you have: ____________________
14. Locate the 1W tap and the Common (Com). The black wire should be the common.
15. Using a wire stripper, strip off about ¼ inch of insulation from the 1W tap and the Common.
16. The remaining transformer taps are not used. If any of them have already been stripped, be sure to clip off the tinned ends to prevent possible shorting or misconnection during installation.

Putting It All Together (see Figure 7–1)

17. Place the tile bridge over the trash can.
18. Drop the back box into the center hole of the tile bridge.
19. Adjust the back box so that the four mounting holes line up with the mounting holes in the tile bridge.
20. Place the prepared ceiling tile, top side up, over the top of the back box. Again, be sure that the marks for the four mounting holes line up.
21. Locate the speaker-grill assembly, and determine the correct orientation for mounting it on the ceiling tile, taking into account the location of the four mounting holes in the speaker grill and the location of the knockout in the back box.
22. Feed the prepared taps of the speaker transformer through the knockout of the back box, and drop the speaker-grill assembly down onto the ceiling tile.
23. Adjust the rotation of the speaker-grill assembly so that the four mounting holes line up with the prepared marks on the ceiling tile.
24. Insert the screws that came with the speaker-grill assembly into the four mounting holes of the grill, and tighten them through. The screws join together the complete assembly, including the speaker-grill assembly, the ceiling tile, the back box, and the tile bridge.
25. The assembly is now complete and ready to be placed in the ceiling grid.

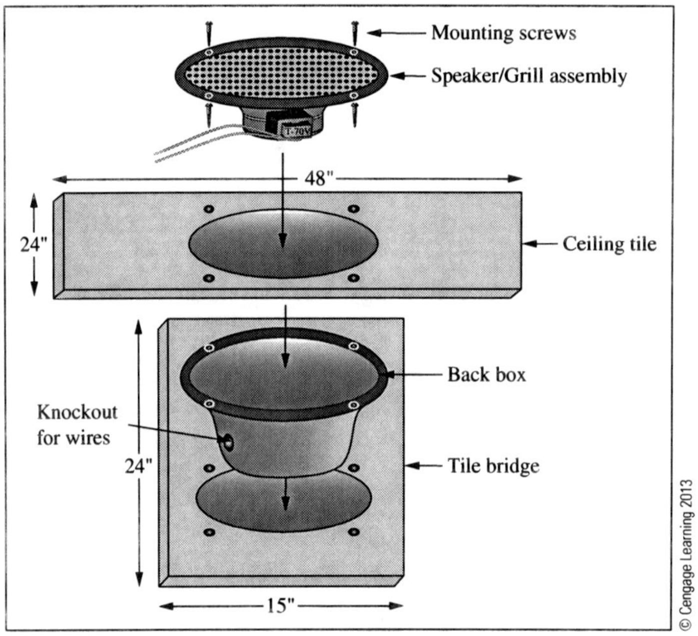

FIGURE 7–1

FINAL SUMMARY

Installation of the back box not only serves to protect the back of the speaker but also improves the frequency response of the output by increasing the pressure on the speaker diaphragm. Increased back pressure helps to improve the forward movement of air molecules and increase overall signal output.

From a wiring point of view, the main concern is that the unused transformer taps do not short together. In addition, always make sure that the size of the speaker, back box, and the tile bridge are compatible with each other, to ensure proper assembly.

FINAL QUESTIONS

1. How does the back box improve the frequency response of the output?
2. Aside from structural support, can you think of another benefit to using the tile bridge?

Name ______________________

Date ______________________

Class ______________________

LAB 8

Class 2 and Class 3 Audio Paging Systems

REQUIRED READING

Introduction to Low Voltage Systems, 2nd edition, Chapter 5

INTRODUCTION

The input of a basic audio paging system includes a paging microphone with muting control and a music source. The output can be configured as either Class 2 or Class 3 and wired accordingly. A Class 3 output utilizes either 70- or 100-volt audio transformers, and Class 2 connects through 25-volt audio transformers.

The wattage maximum for any single Class 2 or Class 3 circuit is limited to 100 watts. Loads exceeding 100 watts must be split between multiple circuits so as not to exceed the listed rating of any CL2 or CL3 cable.

The system volume can be set at the amplifier, or adjusted remotely by a 3-terminal autotransformer. The transformer volume control is wall mounted and locally available within a given space. The voltage rating of the transformer volume control matches the voltage rating of the connecting speakers. The wattage rating of a transformer volume control typically varies between 10 and 100 watts. Always be sure that the total wattage of the connecting speakers does not exceed the rating of the transformer volume control.

A description of how to connect up a basic paging system, as described above, is presented in this lab experiment. Unlike the previous lab experiment, this experiment assembles the speaker-grill assembly directly to the back box. The ceiling tile and the tile bridge are not required.

PRE-JOB QUESTIONS

1. How do you connect an unbalanced, high-impedance microphone to a balanced, low-impedance input?
2. An RCA input is ____________ (balanced or unbalanced) and ________ (low or high) impedance.

OBJECTIVES

- Connect a 70- or 25-volt speaker assembly to the output of an audio amplifier.
- Connect a paging microphone with muting control to the input of an audio amplifier.
- Connect a music source to the input of an audio amplifier.
- Connect a remote transformer volume control to a speaker load.

REQUIRED MATERIALS

12-inch ceiling mount back box

7 feet of 18 AWG, 2-conductor speaker wire

Closed-end crimp connectors

Crimp tool

RCA Y adapter for combining stereo to mono

Stereo headphone adapter to RCA cable

TEST EQUIPMENT

Bogen C20 or C35 amplifier

Shure 522 paging microphone

70-volt or 25-volt, 8-inch ceiling speaker with 13-inch grill cover and mounting screws

Audio input source: CD or MP3 player

70-volt or 25-volt transformer volume control

PROCEDURE

1. Prepare a back box as described in the previous lab exercise.

Connection to the Speaker

2. Cut a piece of 18 AWG, 2-conductor speaker wire to a length of 6 feet, and strip one end.

3. Locate the transformer on the back of the speaker-grill assembly.
4. For this lab, you can use either a 70-volt or 25-volt transformer. Which one are you using? ____________ volt
5. The wires attached to the transformer are marked as follows: Com, 5 W, 2 W, 1 W, ½ W, and ¼ W. Locate the 1-watt tap and the Common. The Common is black.
6. Connect the Com and the 1-watt tap to the stripped end of the speaker wire. Use two closed-end crimp connectors and the crimp tool to make the connections permanent. Be sure to make note of the polarity because it is important to know when you connect to the output of the amplifier.
7. Clip the tinned ends of the unused transformer taps to prevent them from shorting together.
8. Run the speaker cable through the prepared knockout in the back box, and screw the speaker-grill assembly onto the back box.

Connection to the Amplifier

9. Strip the other end of the speaker wire hanging out of the back box.
10. Attach the common side of the speaker wire to the amplifier output labeled COM.
11. Attach the 1-watt tap of the speaker wire to the 70-volt or 25-volt output of the amplifier (as determined by step 4).

Connection of the Microphone

12. In some cases, paging microphones can be configured for either high impedance or low impedance. Before connecting the microphone, be sure that it is configured for low impedance.
13. The microphone cable has five wires to terminate. Three are for the shielded, balanced microphone connection, and two are for the muting control.
14. Identify the microphone input terminals on the back of the amplifier.
15. Connect the microphone shield to the input terminal labeled "GND."
16. Connect the white and green wires to the input terminals labeled "low Z," or "mic." (The white wire represents the high side of the balanced microphone signal, and the green wire, the low side.)
17. Locate the AUX muting control input on the back of the amplifier.
18. Connect the red and black wires to the two terminals labeled "AUX MUTE."
19. Some amplifier inputs are selectable between MIC and AUX. If the amplifier you are using has an input selector switch, be sure that it is set to MIC.

Connection of Music Input

20. The auxiliary input to the amplifier is a single channel mono input. The input connector is an RCA-type connector.
21. Connect a music source from either a portable CD or MP3 player. If you wish to combine the left and right channel, use an RCA Y adapter.

Testing the System

22. Turn on the power to the amplifier and the music source.
23. Adjust the volume control for the music on the auxiliary input.
24. Press the microphone switch, and speak into the microphone.
25. Adjust the microphone volume control to prevent feedback.
26. When you press the microphone switch, the music should automatically cut out or drop to a lower level.
27. Release the microphone switch after making the test page. The music should resume at normal volume level.
28. Shut off the power to the amplifier.

Connection of Remote Transformer Volume Control

29. The voltage of the volume control must match the system you are installing. Be sure the voltage of the transformer volume control matches the voltage of the speaker transformer. __________ volts
30. Disconnect the speaker wire from the amplifier.
31. Cut a 1-foot section of speaker wire, and strip it at both ends.
32. Connect one end of the 1-foot section of speaker wire to the output of the amplifier exactly where the other was previously connected.
33. The transformer volume control has three wires attached to it. Locate the common wire; it should be black.
34. Join together the common of the volume control, the common of the speaker wire, and the common of the short length of speaker wire attached to the amplifier. Use a closed-end crimp and crimping tool to complete the connection.
35. The other two wires connected to the transformer volume control represent the input and the output. The input connects to the amplifier side of the circuit, and the output connects to the speaker side of the circuit. From the documentation that came with your volume control, determine which wire is the input and which is the output. If you do not have documentation, you can determine the input by using an ohm-meter. The resistance between the input and the common lead is fixed and does not change when the volume selector is adjusted between minimum and maximum.
36. Using the closed-end crimp connectors and the crimp tool, connect the input and output as illustrated in Figure 8–1.

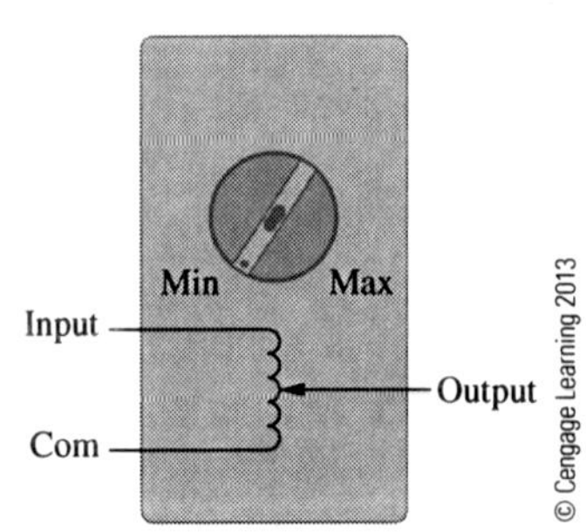

FIGURE 8–1 Transformer volume control.

37. Always be sure that the rated wattage of the volume control is greater than the load connected to it. The wattage rating of the transformer volume control you are using is rated for _____________ watts.
38. Turn on the amplifier and the music source.
39. Test the transformer volume control by adjusting the rotary dial of the control.
40. Turn off all equipment.

FINAL SUMMARY

Figure 8–2 illustrates the wiring of the system described in this lab exercise.

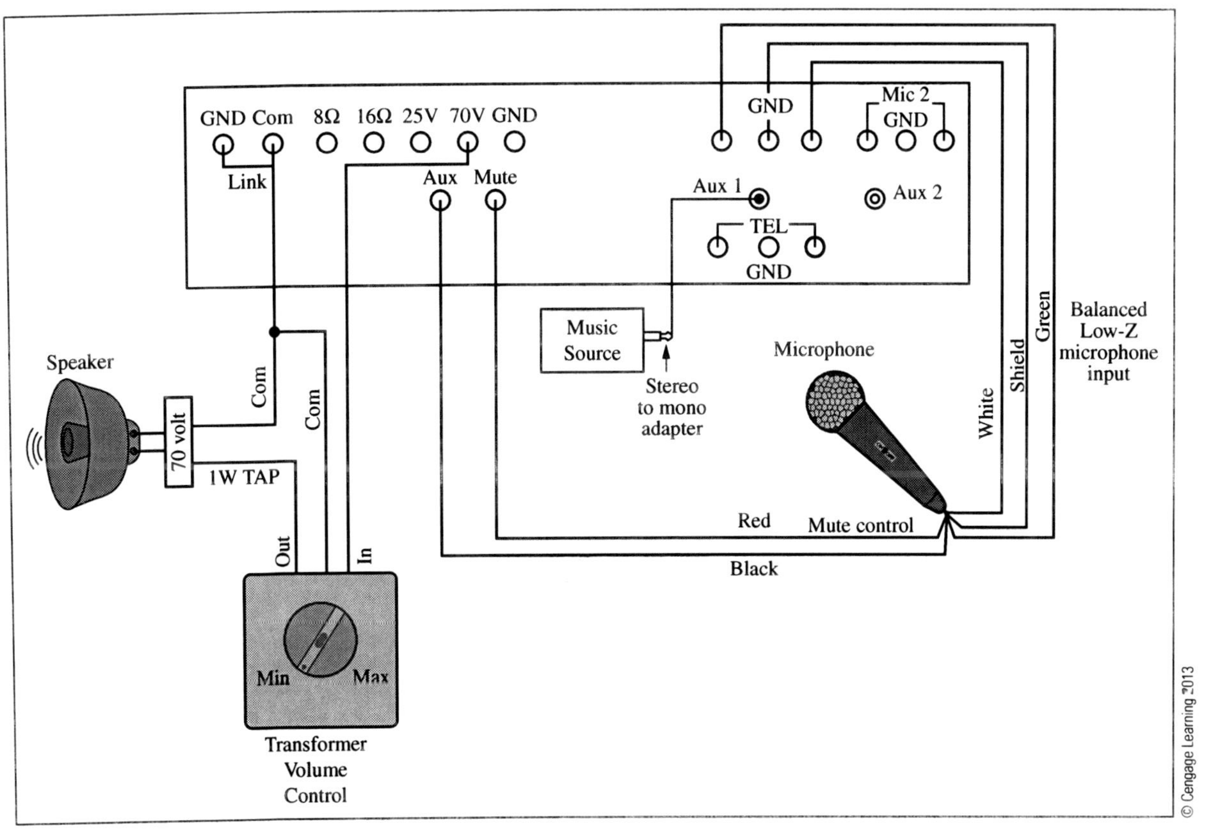

FIGURE 8–2 Connection of audio paging system.

In conclusion, 70-volt, class 3 loads must never be interconnected with 25-volt, class 2 loads, and vice versa. Likewise, 4-, 8-, or 16-ohm loads shall not be used in combination with those of class 2 or class 3 transformers. The intermixing of different load impedances and classifications can potentially overload and damage the amplifier, and it should not be done under any circumstances.

A microphone input shall always be grounded. Never remove the third prong from the power cable or disconnect the equipment ground; to do so could introduce a potential shock hazard. If a hum is present in the output, try implementing an isolated technical ground on all interconnecting equipment.

FINAL QUESTIONS

1. Amplifiers should never be loaded to more than __________ percent of their rated output.
2. A telephone paging input is line level, balance, and __________ ohms.

Name ____________________

Date ____________________

Class ____________________

LAB

9

Network Cables and Connectors

REQUIRED READING

Introduction to Low Voltage Systems, 2nd edition, Chapter 6

INTRODUCTION

Computer networks are most commonly connected through 4-pair UTP cables. Terminations typically use 8P8C, modular RG-45 connectors. This lab discusses the procedure for attaching such connectors, as well as the differences between the T568A and T568B wiring standards.

A straight-through cable is made by matching the order of the colored wires on both ends of the cable. The order of the wires is determined by the wiring standard of the equipment and connecting devices. To demonstrate the differences between the standards, you build a straight-through cable of each variety and test the cable for continuity and correct order of pairs.

PRE-JOB QUESTIONS

1. How many pins does an RJ-45 connector have?
2. What does the "RJ" stand for?
3. How does an RJ-14 connector compare to an RJ-45?
4. How many pairs are in a CAT-5e cable, and what are their colors?

OBJECTIVES

- Build a straight-through network cable, using the T568A standard.
- Build a straight-through network cable, using the T568B standard.
- Test the cable for continuity and swapped pairs.

REQUIRED MATERIALS

2 feet of CAT-5e cable
8P8C RJ-45 modular plug crimp tool
8P8C RJ-45 modular plug connectors
Ruler

TEST EQUIPMENT

Ideal LinkMaster Pro XL cable tester

PROCEDURE

Terminating a T568A Cable

1. Cut a 1-foot section of CAT-5e cable.
2. Remove 1 inch of the outer jacket from one end of the cable.
3. To accommodate the T568A standard, arrange the colored strands, from left to right, in the following order:
   White\Green
   Green
   White\Orange
   Blue
   White\Blue
   Orange
   White\Brown
   Brown
4. Trim the ends of the wires with the cutting edge of the crimp tool or a wire cutter, so that they are equal in length. After trimming, they should measure not more than ½ inch in length.
5. Insert the wires into the modular RJ-45 connector, in the order shown in Figure 9–1.

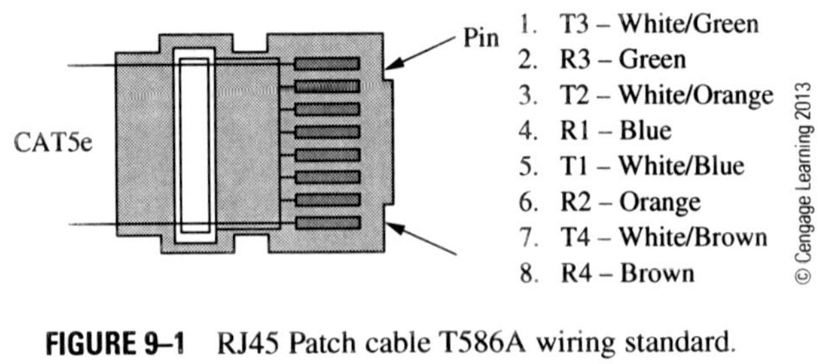

FIGURE 9–1 RJ45 Patch cable T586A wiring standard.

6. Slide the RJ-45 connector, with the latch down, into the position marked "8P" on the crimp tool. The connector clicks into place when the latch engages the die.
7. Squeeze the crimp tool as far as it will go, and then release.
8. Repeat steps 2–7 on the other end of the cable.
9. Insert both ends of the cable into the cable tester to verify that the cable is properly constructed.
10. If the cable fails the test, cut off each connector and try again.

Terminating a T568B Cable

1. Cut a 1-foot section of CAT-5e cable.
2. Remove 1 inch of the outer jacket from one end of the cable.
3. To accommodate the T568B standard, arrange the colored strands, from left to right, in the following order:

   White\Orange
   Orange
   White\Green
   Blue
   White\Blue
   Green
   White\Brown
   Brown
4. Trim the ends of the wires with the cutting edge of the crimp tool or a wire cutter, so that they are equal in length. After trimming, they should measure not more than ½ inch in length.
5. Insert the wires into the modular RJ-45 connector, in the order shown in Figure 9–2.

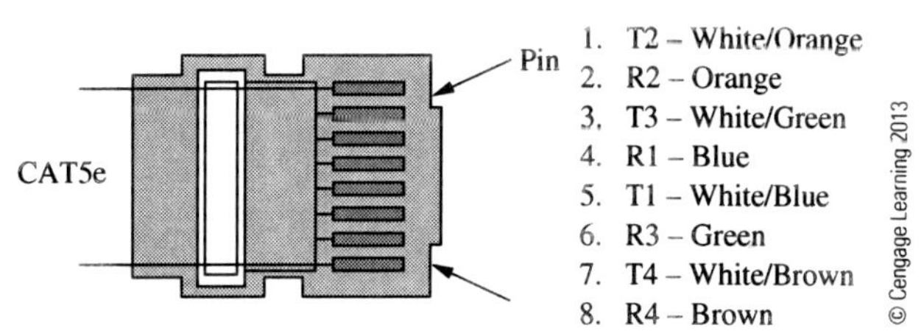

FIGURE 9–2 RJ45 Patch cable T586B wiring standard.

6. Slide the RJ-45 connector, with the latch down, into the position marked "8P" on the crimp tool. The connector clicks into place when the latch engages the die.
7. Squeeze the crimp tool as far as it will go, and then release.
8. Repeat steps 2–7 on the other end of the cable.

9. Insert both ends of the cable into the cable tester to verify that the cable is properly constructed.
10. If the cable fails the test, cut off each connector and try again.

FINAL SUMMARY

The difference between the T568A standard and the T568B standard is that the orange pairs and the green pairs are swapped. Figure 9–3 shows the pair positions and the color codes for the T568A and T568B standards.

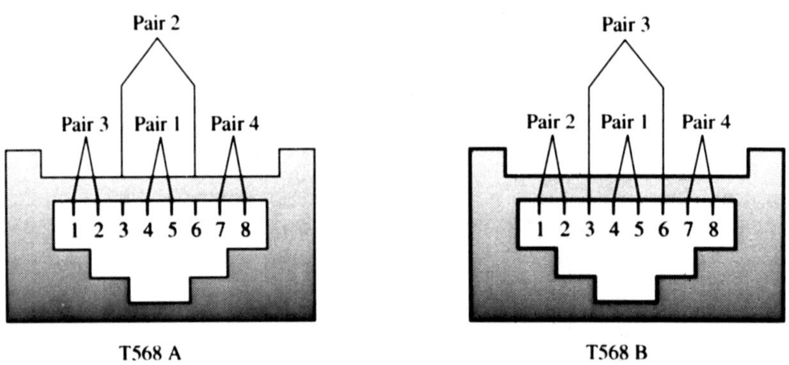

T568 A	T568 B
1. T3 – White/Green	1. T2 – White/Orange
2. R3 – Green	2. R2 – Orange
3. T2 – White/Orange	3. T3 – White/Green
4. R1 – Blue	4. R1 – Blue
5. T1 – White/Blue	5. T1 – White/Blue
6. R2 – Orange	6. R3 – Green
7. T4 – White/Brown	7. T4 – White/Brown
8. R4 – Brown	8. R4 – Brown

FIGURE 9–3 EIA/TIA T586A & B wiring standard.

The T568A standard is used in residential installations and for analog voice applications. Pairs 1 and 2 are wired the same as an RJ-14 two-line residential phone jack.

The T568B standard is more commonly used in commercial installations using multiline electronic key phone systems, ISDN, T1, and for most data applications. Incidentally, the T568B standard was originally used by AT&T for the Merlin phone system. To remember, just associate "B" with Bell, as in Bell Telephone.

FINAL QUESTIONS

1. When installing CAT-5 or CAT-5e cable, it is recommended that not more than _____ inch of the individual pairs are untwisted.
2. Which wiring standard is most commonly used in residential installations?
3. Which wiring standard is most commonly used on T1 lines?
4. Which wiring standard matches to Pairs 1 and 2 of an RJ-14 connector?

Name ______________________

Date ______________________

Class ______________________

LAB 10

Crossover Cable

REQUIRED READING

Introduction to Low Voltage Systems, 2nd edition, Chapter 6

INTRODUCTION

The Crossover Cable

In some cases, there is a need to connect equipment and devices between two different wiring standards. To do so, a crossover cable is required. A crossover cable is made by configuring one end as the T568A wiring standard and the other end as T568B. By doing so, the positions of Pairs 2 and 3 within the connecting cable are swapped.

A Reversed Patch Cable

Another type of cable is called a reversed patch cable. To build a reversed RJ-45 patch cable, pin 1 is wired to pin 8; pin 2 is wired to pin 7; pin 3 is wired to pin 6; and so forth. Manufacturers build their equipment to require either a straight-through cable, as described in the previous lab exercise, or a reversed patch cable, depending on the wiring specification of the system. Reversed patch cables are most often seen on analog voice telephone systems. To know which type to use, always consult the manufacturer's instructions.

This lab exercise has you build and test a crossover cable.

PRE-JOB QUESTIONS

1. For the RJ-45 Jack shown in Figure 10–1, indicate the pair orders and color codes for the T568A wiring standard.

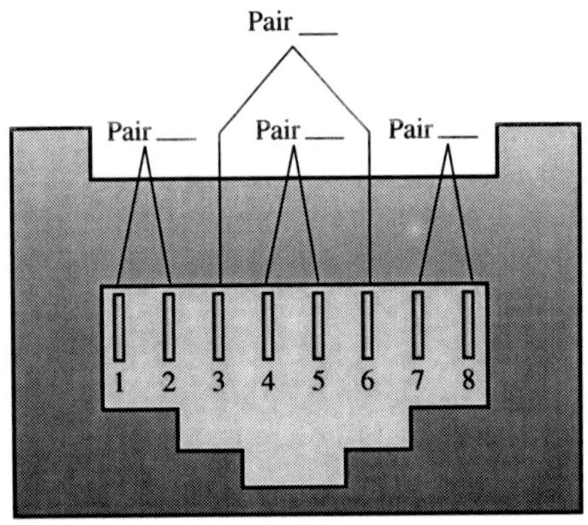

Pin	Color
1.	
2.	
3.	
4.	
5.	
6.	
7.	
8.	

FIGURE 10–1 T586A wiring standard.

For the RJ-45 Jack shown in Figure 10–2, indicate the pair orders and color codes for the T568B wiring standard.

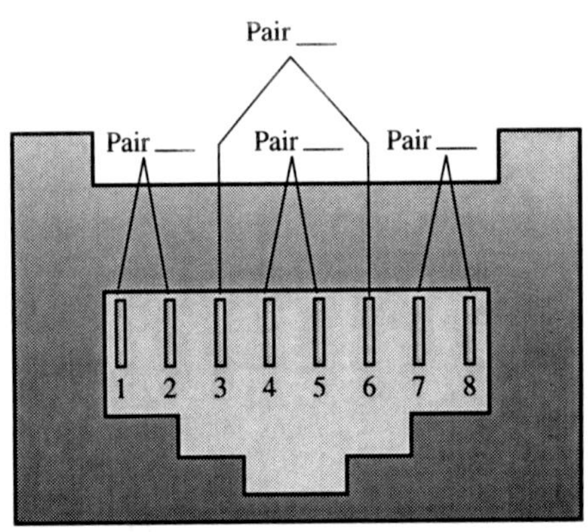

Pin	Color
1.	
2.	
3.	
4.	
5.	
6.	
7.	
8.	

FIGURE 10–2 T586B wiring standard.

OBJECTIVES

- Build a crossover cable.
- Test the cable for continuity and swapped pairs.
- Determine the wiring pin-out of a reversed patch cable.

REQUIRED MATERIALS

CAT-5 cable
Crimp tool
RJ-45 connectors
Ruler

TEST EQUIPMENT

Ideal, LinkMaster Pro XL cable tester

PROCEDURE

Terminating a Crossover Cable

1. Cut a 1-foot section of CAT-5e cable.
2. Remove 1 inch of the outer jacket from one end of the cable.
3. To accommodate the T568A standard, arrange the colored strands, from left to right, in the following order:

   White\Green
   Green
   White\Orange
   Blue
   White\Blue
   Orange
   White\Brown
   Brown
4. Trim the ends of the wires with the cutting edge of the crimp tool or a wire cutter so that they are equal in length. After trimming, they should measure not more than ½ inch in length.
5. Insert the wires into the modular RJ-45 connector, in the order shown in Figure 10–3.

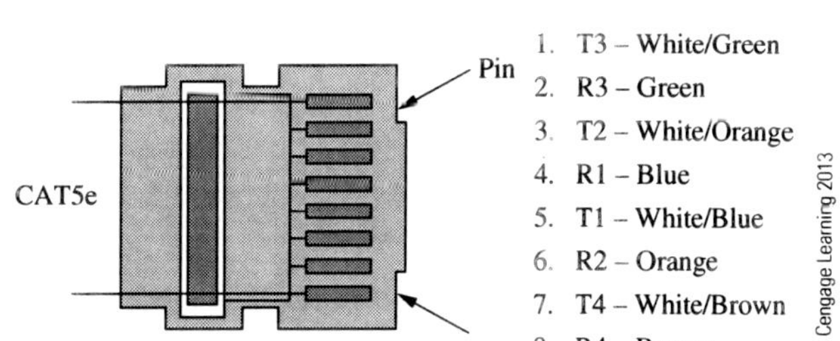

FIGURE 10–3 RJ45 Patch cable T586A wiring standard.

6. Slide the RJ-45 connector, with the latch down, into the position marked "8P" on the crimp tool. The connector clicks into place when the latch engages the die.
7. Squeeze the crimp tool as far as it will go, and then release.
8. Remove 1 inch of the outer jacket from the other end of the cable.

9. To accommodate the T568B standard, arrange the colored strands, from left to right, in the following order:

   White\Orange
   Orange
   White\Green
   Blue
   White\Blue
   Green
   White\Brown
   Brown

10. Trim the ends of the wires with the cutting edge of the crimp tool or a wire cutter so that they are equal in length. After trimming, they should measure not more than ½ inch in length.
11. Insert the wires into the modular RJ-45 connector in the order shown in Figure 10–4.

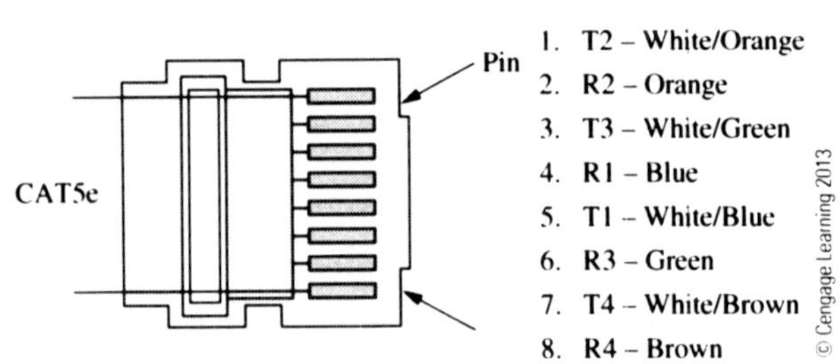

FIGURE 10–4 RJ45 Patch cable T586B wiring standard.

12. Slide the RJ-45 connector, with the latch down, into the position marked "8P" on the crimp tool. The connector clicks into place when the latch engages the die.
13. Squeeze the crimp tool as far as it will go, and then release.
14. Insert both ends of the cable into the cable tester to verify that the cable is properly constructed. The cable tester should read "crossover," indicating that Pairs 2 and 3 are swapped at each end of the cable.
15. If the cable fails the test, cut off each connector and try again.

FINAL SUMMARY

Crossover cables are used when connecting between two different wiring standards. In most cases, crossover cables are made to connect between the T568A and T568B wiring standards.

Reversed patch cables are required by some manufacturers and are indicated in the installation manual for the equipment. Equipment wiring diagrams always indicate what type of wiring standard is required and whether the cables are to be straight-through, reversed, or crossover.

FINAL QUESTIONS

1. A patch cable can be made in one of three varieties, straight-through, _________, or _________.
2. Where are crossover cables required?
3. Where are reversed patch cables required?
4. Draw how you would wire a reversed patch cable between two RJ-45 connectors shown in Figure 10–5.

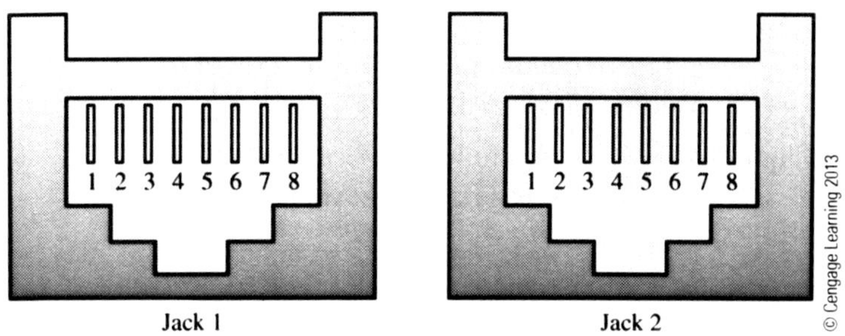

FIGURE 10–5

Name ____________________

Date ____________________

Class ____________________

LAB 11

Half-Wave Rectifiers and LEDs

REQUIRED READING

Introduction to Low Voltage Systems, 2nd edition, Chapter 7

INTRODUCTION

A rectifier converts alternating current (AC) to pulsating direct current (DC). The electronic component that rectifies an AC signal is a diode. A diode is a semiconductor device that allows current to flow in only one direction, acting much like an electronic switch (see Figure 11–1). Current flow through a diode is from cathode to anode. The voltage of a forward biased silicon diode is between 0.6 and 0.7 volt. For a germanium diode, the voltage is 0.3 volt. When a diode turns on, in forward bias mode the current must be limited through a series resistor, or the diode will burn up. The manufacturer specification data sheet for a diode lists the maximum forward current flow for the device as I_{FM}. A series resistor is used to limit the forward current flow to less than the forward maximum.

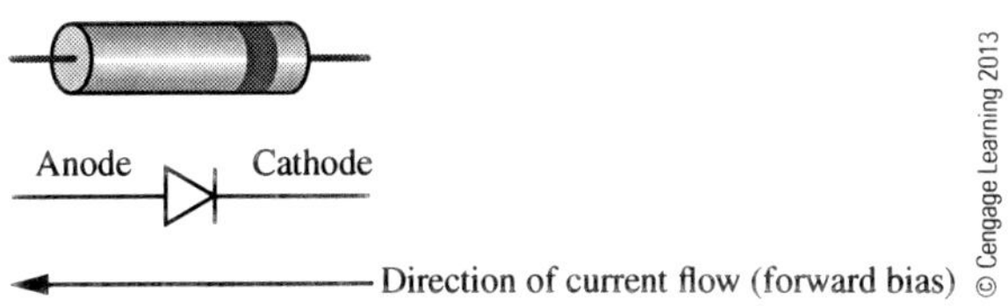

FIGURE 11–1 A Diode

A diode is reverse biased when the current flow is from anode to cathode. In reverse bias mode, the diode shuts off and acts like an open switch. When a diode is reverse biased, it is important that the peak voltage of the input ac voltage is not greater than the maximum peak reverse voltage (V_{PRV}), as specified on the manufacturer's specification data sheet. Connecting a diode to a peak voltage higher than the V_{PRV} destroys the diode.

A light-emitting diode (LED) operates like a rectifier diode except that the forward biased voltage drop of an LED is between 1.3 and 1.7 volts, depending on the level of current flow through the diode. But unlike rectifier diodes, which convert AC to DC, LEDs are only used in DC circuits as indicator lamps. The schematic symbol for an LED is shown in Figure 11–2. The minimum required current flow to light an LED is typically 20 milliamperes (mA). Maximum power dissipation for most LEDs is less than 180 milliwatts (mW), which equates to about 120 mA of current for a 1.5-volt LED. An LED also requires a series resistor to limit the current flow to the desired value. Connecting any diode in a circuit without a series-limiting resistor destroys the diode. The resistor is also required to drop the extra circuit voltage, over and above the forward bias voltage of the diode.

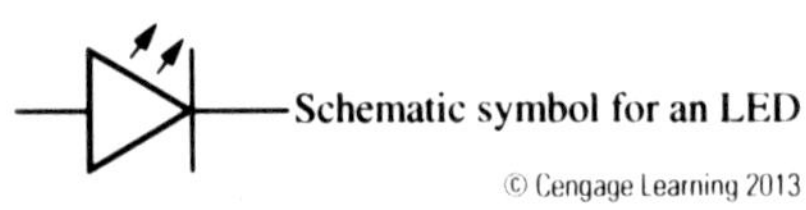

FIGURE 11–2 Schematic symbol for an LED

PRE-JOB QUESTIONS

1. What is the definition of a rectifier?
2. What component provides rectification?
3. Convert 12 volts AC, RMS to a peak voltage.
4. What is the voltage of a forward biased silicon diode?
5. For a half-wave rectifier, the load current flows for how many degrees of the input cycle?
6. Draw the output waveform for a half-wave rectifier.
7. Is the output of a half-wave rectifier pure DC? Explain.
8. What is the AC component of a half-wave rectifier called?
9. What is the typical voltage measured across a forward bias red LED?
10. Draw in the direction of current flow for the circuit shown in Figure 11–3.

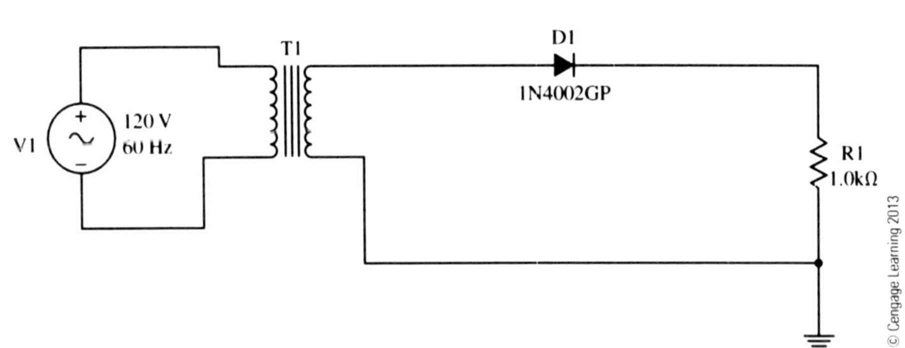

FIGURE 11–3

OBJECTIVES

- Test, measure, and analyze a half-wave rectifier circuit.
- Calculate the expected output voltages of a half-wave rectifier.
- Test, measure, and analyze a half-wave rectifier circuit with an LED load.
- Explain the difference between positive DC voltage and negative DC voltage.

REQUIRED MATERIALS

1N4002 diode (with specification sheets)

1-kiloohm, ½-watt resistor

Red LED (with specification sheets)

TEST EQUIPMENT

PAD-234 analog digital trainer

Digital multimeter (DMM)

PROCEDURE

Half-Wave Rectifier

1. Connect the circuit on the PAD-234 trainer, as shown in Figure 11–4. The white or black band on the diode indicates the cathode side of the component.

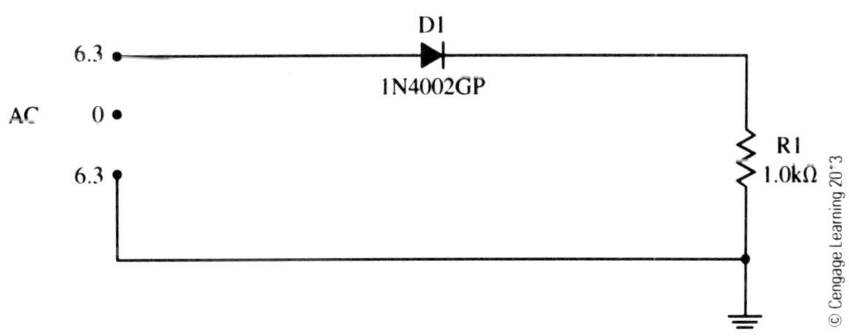

FIGURE 11–4

2. Measure the Input AC voltage with the DMM. Record the value in Table 11–1 as Input AC_{RMS}.
3. Measure the AC voltage across R1 with the DMM. Record the value in Table 11–1 as AC ripple.
4. Switch the DMM to DC and measure the DC voltage at R1. Be sure to place the meter common lead at the bottom of the circuit. Record the value in Table 11–1 as DC output.

5. Turn off power.
6. Insert a red LED (Figure 11–5) in series with the load, as shown in Figure 11–6.

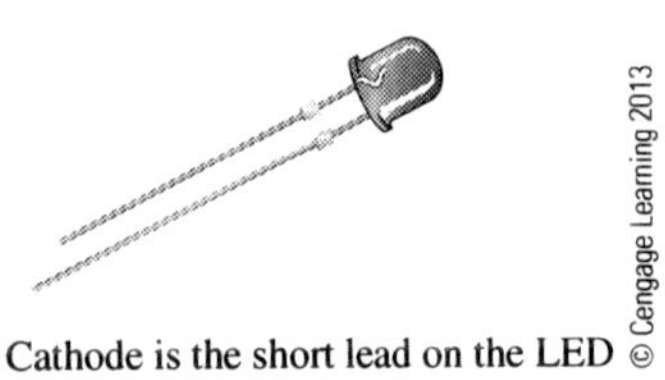

**The Cathode is the short lead on the LED

FIGURE 11–5

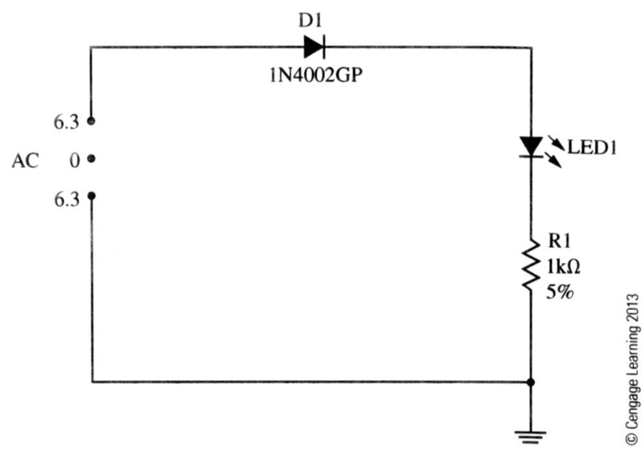

FIGURE 11–6

7. Turn on the power. Does the LED light? YES☐ NO☐
8. Measure the DC voltage of R1 again. Be sure to place the meter common lead at the bottom of the circuit. Record the value in Table 11–1 as LED resistor voltage.
9. Measure the DC voltage across the LED and record the value in Table 11–1 as LED V_{FWD}. Be sure to place the common lead of the DMM at the cathode when taking this measurement.
10. Turn off the power and connect the circuit shown in Figure 11–7 with the 1N4002 diode reversed.

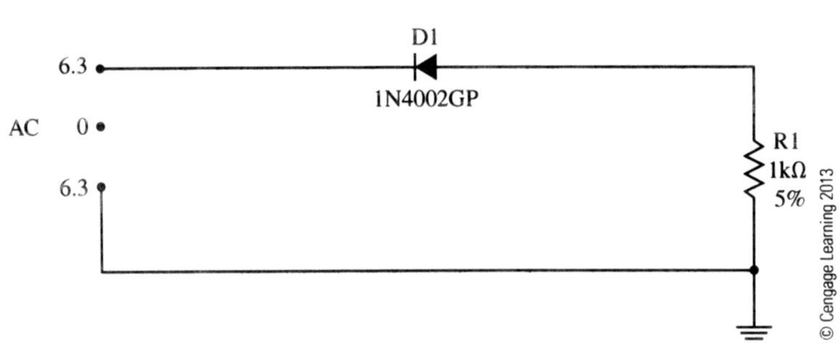

FIGURE 11–7

11. Measure the DC voltage of R1 again. Be sure to place the meter common lead at the circuit ground. Record the value in Table 11–1 as DC_{REV} output.
12. Turn off the power and connect the LED, as shown in Figure 11–8.

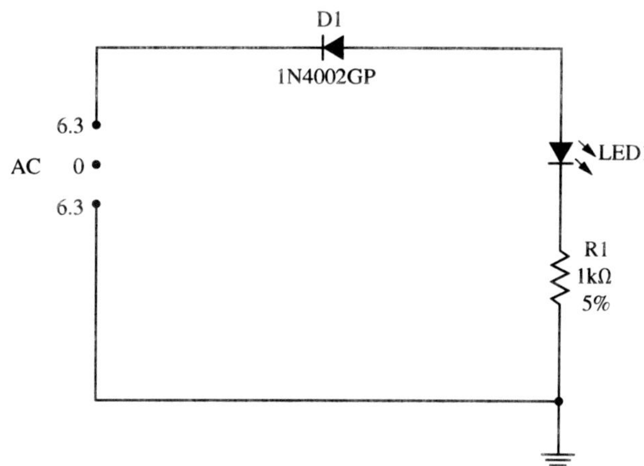

FIGURE 11–8

13. Turn on the power. Does the LED light? YES☐ NO☐
14. Measure the DC voltage across the LED.
    Record the measurement in Table 11–1 as LED voltage REV.
15. Turn off the power, and then turn the LED around.
16. Turn on the power. Does the LED light? YES☐ NO☐

TABLE 11–1 Half-Wave

Voltage	Measured	Calculated
Input AC_{RMS}		
Input AC peak		
Peak load voltage		
AC ripple		
DC output		
LED voltage FWD		
LED resistor voltage		
DC_{REV} output		
LED voltage REV		
Ripple %		

Half-Wave Rectifier Calculations

1. Calculate the Input AC peak from the measured Input AC_{RMS}, and record the value in Table 11–1.

$$ACV_{Peak} = V_{RMS}/0.707$$

2. The peak load voltage is then calculated by subtracting a diode drop.

$$VL_{Peak} = ACV_{Peak} - 0.7 \text{ volt}$$

3. The AC ripple at the load is calculated using the following formula:

$$VL_{Ripple} = VL_{Peak} \times 0.385$$

4. The DC load voltage is calculated using the following formula:

$$VL_{DC} = \frac{VL_{peak}}{\pi}$$

5. AC ripple % is calculated using the following formula:

$$Ripple\% = \frac{VL_{Ripple}}{VL_{DC}} \times 100$$

6. Calculate all values using the above formulas and enter the results in Table 11–1.

DC Average

A half-wave rectifier allows load current to flow for only the first 180 degrees of the input AC cycle. This means that the diode is OFF every half-cycle.

If you average the instantaneous voltages around a sine wave, from zero to 360 degrees, the result would be zero because the negative cycle cancels out or nulls the positive cycle. Simply stated,

$$+V_{Peak} + (-V_{Peak}) = 0$$

So, how then can a sine wave average be calculated? The answer is simple: by removing half of the output signal. That's what the diode does. For the circuit in Figure 11–4, the diode only conducts for 180 degrees, during the positive cycle, thereby removing the negative cycle.

The DC average for a half-wave rectifier is then calculated by averaging the voltages from zero to 180 degrees. Mathematically, this is equal to

$$VL_{DC} = \frac{VL_{peak}}{\pi}$$

VL_{Peak} represents the maximum voltage at 90 degrees, and π represents 180 degrees of a circular rotation, as the signal transitions from zero to maximum and back to zero again.

Ripple

A half-wave rectified output is a composite waveform, consisting of both AC and DC mixed together in one signal. A half-wave output is therefore not pure DC. The constant fluctuation of the half-wave output voltage results in an AC noise component referred to as *ripple*. This is not a small problem. The ripple is very significant, equal to 38.5% of the peak voltage. In reality, there is more AC ripple in the output than DC, which is why the ripple % for a half-wave rectifier is 121%, meaning that there is 21% more AC noise in the signal than actual DC.

For some circuits, 121% ripple is not a problem, but for sensitive electronic computer circuits or for audio circuits, this amount of ripple would be a disaster.

In the case of audio circuits, you can actually hear the AC ripple buzzing through the speakers of the amplifier as an obnoxious interfering noise.

To improve the DC output and remove the unwanted ripple, a power supply filter and voltage regulator are required.

Negative DC

Negative DC is produced by turning the rectifier diode around, causing the current to flow in the opposite direction. Figure 11–7 has the diode reversed, and the current is now flowing into ground. In this case, the ground is now the more positive point in the circuit, which is why the voltage reading at the load measures negative.

LED

An LED lights only if it is forward bias, as in Figure 11–6. For Figure 11–8, the LED is off because the negative DC voltage reverse biases the LED. Turning the LED around allows it to be forward biased again, and the light turns back on.

A red LED typically drops between 1.3 and 1.7 volts DC. Always check the manufacturer's specification data on the type of LED that you are using. Exceeding the maximum forward current destroys the LED.

FINAL QUESTIONS

1. For a half-wave rectifier circuit, what determines the direction of current flow?
2. Looking at the circuit in Figure 11–4, if the diode shorted, what would happen to the output?
3. Looking at the circuit in Figure 11–4, if the diode opened, what would happen to the output?
4. What voltage did you measure across the LED when it was forward biased? ________ volts.
5. How did the load voltage change when the rectifier diode was reversed in Figure 11–7?
6. What is the definition of a *composite waveform*?
7. How can you test for ripple at the output if you only have a DMM?
8. A negative DC half-wave rectifier causes current to flow ________ (out of, or into) the ground.
9. Based on the specifications for the 1N4002 diode, what is the maximum allowable current flow through the load?
10. Based on the specifications for the 1N4002 diode, what is the maximum allowable peak input voltage?

Name ______________________

Date ______________________

Class ______________________

LAB 12 Conduit Fill

REQUIRED READING

Introduction to Low Voltage Systems, 2nd edition, Chapter 8

INTRODUCTION

The starting point for conduit fill is *NEC Article 300.17, Number and Size of Conductors in Raceway.*

> ***300.17 Number and Size of Conductors in Raceway***
> *The number and size of conductors in a raceway shall not be more than will permit the dissipation of head and ready installation or withdrawal of the conductors without damage to the conductors or to their insulation.*
>
> *Informational Note: See the following sections of this Code:*
>
> *intermediate metal conduit, 342.22;*
> *rigid metal conduit, 344.22;*
> *flexible metal conduit 348.22;*
> *liquidtight flexible metal conduit, 350.22;*
> *rigid nonmetallic conduit, 352.22;*
> *liquidtight nonmetallic flexible conduit, 356.22;*
> *electrical metallic tubing, 358.22;*
> *flexible metallic tubing, 360.22;*
> *electrical nonmetallic tubing, 362.22;*
> *cellular concrete floor raceways, 372.11;*
> *cellular metal floor raceways, 374.5;*
> *metal wireways, 376.22;*
> *surface nonmetallic raceways, 388.22;*
> *underfloor raceways, 390.5;*
> *fixture wire, 402.7;*

theaters, 520.6;
signs, 600.31(C);
elevators, 620.33;
audio signal processing, amplification, and recording equipment, 640.23(A) and 640.24;
Class1, Class 2, and Class 3 circuits, Article 725;
fire alarm circuits, Article 760;
and optical fiber cables and raceways, Article 770.

All of the above referenced informational notes require that conduit fill restrictions be followed. Notice that *Article 800* is not referenced.

NEC 800.110, Raceways for Communications Wires and Cables.

(A)(1) Raceways Recognized in Chapter 3
Communications wires and cables shall be permitted to be installed in any raceway included in Chapter 3. The raceway shall be installed in accordance with the requirements of Chapter 3.

(A)(2) Other Permitted Raceways
Communications wires and cables shall be permitted to be installed in listed plenum communications raceway, listed riser communications raceways, or listed general-purpose communications raceway (800.182) selected in accordance with the provisions of 800.113, and installed in accordance with 362.24 through 362.56, where the requirements applicable to electrical nonmetallic tubing apply.

(B) Raceways Fill for Communications Wires and Cables
The raceway fill requirements for Chapters 3 and 9 shall not apply to communications wires and cables.

As stated above, *Article 800.110(A)* only requires that *Chapter 3* be followed where installing raceways, but *(B)* states that conduit fill restrictions shall not apply to communications wires and cables. But is this a good design practice? Remember, the *NEC* is a safety manual and not intended as a design specification. Safety involves the protection of persons and property from hazards. A design specification is different; it not only addresses safety but also includes additional requirements to guarantee maximum system performance.

Keep in mind, the two main reasons for conduit fill restriction are to allow ample room inside the pipe to dissipate heat and to prevent damage to conductors when pulling them in and out of the conduit. Although communication cables do not dissipate any noticeable heat to be concerned about, it is still important to follow conduit fill restrictions so as not to damage the conductors when pulling them through overfilled spaces.

The industry standard for maximum pull tension on a single CAT5e cable is to be no more than 25 lbs per cable. Pull tensions greater than 25 lbs could possibly damage or destroy the characteristic impedance of the cable as a result of it being stretched or pulled too tightly. In such cases, although the damage may not be visible, it can be detected through poor performance on high-frequency systems; a cable originally intended to transmit at 100 Mbps now barely transmits at 1 Mbps.

According to *Chapter 9, Table 1* of the *NEC*, 40% is the recommended conduit fill restriction for communications cables where there are more than two cables installed in a conduit.

Table 1, Chapter 9 of the *NEC* also has additional notes that we need to look at:

> *Note 3: Equipment grounding and bonding conductors, where installed in conduits, shall be included when calculating conduit or tubing fill. The actual dimensions of bare or insulated conductors shall be used.*
>
> *Note 4: Where conduit or tubing nipples having a maximum length not to exceed 24 inches (600 mm) and are installed between boxes, cabinets, and similar enclosures, the fill shall be permitted to be 60 %. Temperature adjustments of 310.15 (B) shall not apply.*
>
> *Note 5: For conductors not included in Chapter 9, such as multiconductor cables and optical fiber cables, the actual dimensions shall be used.*

In such a case, you can either look up the actual dimension in a specification data sheet or calculate it yourself from the geometric formula for area:

$$\text{Area} = \pi r^2 \text{ and } r = \text{diameter}/2$$

> *Note 8: Where bare conductors are permitted by other sections of the Code, the dimensions for bare conductors in Table 8, Chapter 9 of the NEC shall be permitted.*
>
> *Note 9: A multiconductor cable or flexible cord of two or more conductors shall be treated as a single conductor for calculating percentage conduit fill area. For cables that have elliptical cross sections, the cross-sectional area calculation shall be based on using the major diameter of the ellipse as a circle diameter.*

PRE-JOB QUESTIONS

1. List two reasons why conduit fill restrictions are important.
2. Give the *NEC* code reference as to where conduit fill is found in the *Code* book.
3. If a cable is not included in the fill tables, how do you calculate the cross-sectional area?
4. Complete the percent of cross section in the table below, as found in *Table 1, Chapter 9* of the 2011 *NEC*.

Number of Conductors	Percent Fill
1	
2	
Over 2	

OBJECTIVES

- Calculate and determine the required conduit size for an installation, using the *NEC* fill tables.
- Calculate and determine the required conduit size for conductors installed in all types of raceways, as covered in *Chapter 3* of the *NEC*.
- Calculate the cross-sectional area of a conductor through the use of geometry, and determine the required conduit size needed for an installation.
- Calculate the cross-sectional area of an elliptical cable.

REQUIRED MATERIALS

2011 *National Electrical Code Book*
Calculator

PROCEDURE

Using the 20011 National Electrical Code, *follow the steps below and fill in the blanks to answer the following question:*

1. What size flexible metal conduit is needed for twenty-five 18 AWG XFF and twenty-five 16 AWG XFF conductors?
2. Look up the article reference for flexible metal conduit in the informational note of *NEC 300.17*, and turn to that section.
   Go to *flexible metal conduit, Article* ____________.
3. Read the reference titled *Number of Conductors*, and proceed to *Table* __________, *Chapter* __________.
4. What percentage of fill is used for this example?
   __________%
5. Read *Note 6*, in *Table 1, Chapter 9* of the *NEC*.
6. Turn to *Table 5, Chapter 9* of the *NEC* to look up the cross-sectional area of an 18 AWG XFF conductor.
   Dimension of 18 AWG XFF = ______________ in.^{2}
7. Multiply the dimension of the 18 AWG XFF by 25.
   Total dimension of twenty-five 18 AWG conductors: ______________ in.^{2}
8. From *Chapter 9, Table 5* of the *NEC*, look up the cross-sectional area of a 16 AWG XFF conductor.
   Dimension of 16 AWG XFF = ______________ in.^{2}
9. Multiply the dimension of 16 AWG XFF by 25.
   Total dimension of twenty-five 18 AWG conductors: ______________ in.^{2}
10. Add the values obtained in steps 7 and 9 to obtain the total cross-sectional area of all 50 conductors.
    Total dimension of all 50 conductors ______________ in.^{2}

11. Turn to *Table 4* in *Chapter 9* of the *NEC*, and find the table referencing flexible metal conduit.
12. Reading from the last column "Over 2 conductors, 40%," determine the size of conduit needed to fit the total dimension of all 50 conductors, as calculated in step 10.
    ______________-inch FMC
13. Recalculate the above problem if there are sixty16 AWG XFF. Show your work below.
    ______________ inch FMC
14. What size intermediate metal conduit is needed for ten 16 AWG TFF and twenty-five 18 AWG SFF-1 conductors? Show your work.
    ______________-inch IMC
15. Recalculate problem 13 for only one TFF and one SFF-1 conductor. Show your work.
    ______________-inch IMC

FINAL SUMMARY

Conduit fill is concerned with two primary issues: heat dissipation and possible damage to insulation as conductors and cables are being installed. All installers need to know how to calculate conduit fill. The informational note for *NEC 300.17* references all sections of the *Code* that require conduit fill restriction. Although *300.17* does not reference *Article 800*, and *Section 800.110* specifically exempts conduit fill restrictions for communications wires and cables, it is still an acceptable industry standard to use a 40% fill restriction where installing communications cables in conduits. Although excessive heat dissipation is not a concern for communications cables, possible damage to the conductors during installation is, and care should always be taken so as not to destroy the characteristic impedance of the cable, which can directly affect system performance.

FINAL QUESTIONS

1. What size electrical nonmetallic tubing (ENT) is needed for 25 CAT5e cables? Use a dimension of 0.25 inch for the diameter of a single CAT5e cable. See *Note 5* from *Table 1, Chapter 9* of the 2011 *NEC*, on page 711.
2. What size intermediate metal conduit (IMC) is needed for two satellite cables having an elliptical dimension of 0.25 inch × 0.55 inch? Also include a bare 6 AWG solid copper bonding conductor in the calculation. See *Notes 8 and 9* from *Table 1, Chapter 9* of the 2011 *NEC*, on page 711 of this chapter.

Name ____________________

Date ____________________

Class ____________________

LAB 13

Fire Alarm Systems

REQUIRED READING

Introduction to Low Voltage Systems, 2nd edition, Chapter 9

REFERENCES

Silent Knight 5802 user manuals

User manuals for:

ESL 449CT	4-wire smoke detector
ESL 429CT	2-wire smoke detector
FED 4050-001T-201	Pull station
WHE RSS24MCW-FR	Strobe lights
System Sensor SpectrAlert	H12/24 horn
EDW 1508-AQN5	Magnetic door holder

INTRODUCTION

It is recognized that an understanding of system devices, terminations, panel connections, and basic troubleshooting techniques are necessary skills for all limited energy installers. Understanding the theory of a fire alarm control panel (FACP) is greatly enhanced by having a working panel along with input and output devices for students to examine, connect, test, and troubleshoot. The upcoming lab procedure outlines the connections, function, and testing of a basic system.

It is not the intent of this exercise to teach a specific type of fire alarm control panel. Although the type of panel used in this exercise is the Silent Knight 5802, any type of basic fire alarm control panel can be used for this training exercise.

System programming is not covered in the upcoming lab procedure. Therefore, it is the responsibility of the instructor to perform the initial system setup and programming before attempting the exercise. Any unique manufacturer specifics must also be addressed by the instructor to ensure that the system functions as required.

Before doing this lab exercise, it is recommended that the control panel be mounted on a piece of ¾-inch plywood and that all internal panel connections be wired to external terminal strips, as shown in Figure 13–1, and labeled. To do so helps prevent any undo wear and tear on the internal connections of the control panel.

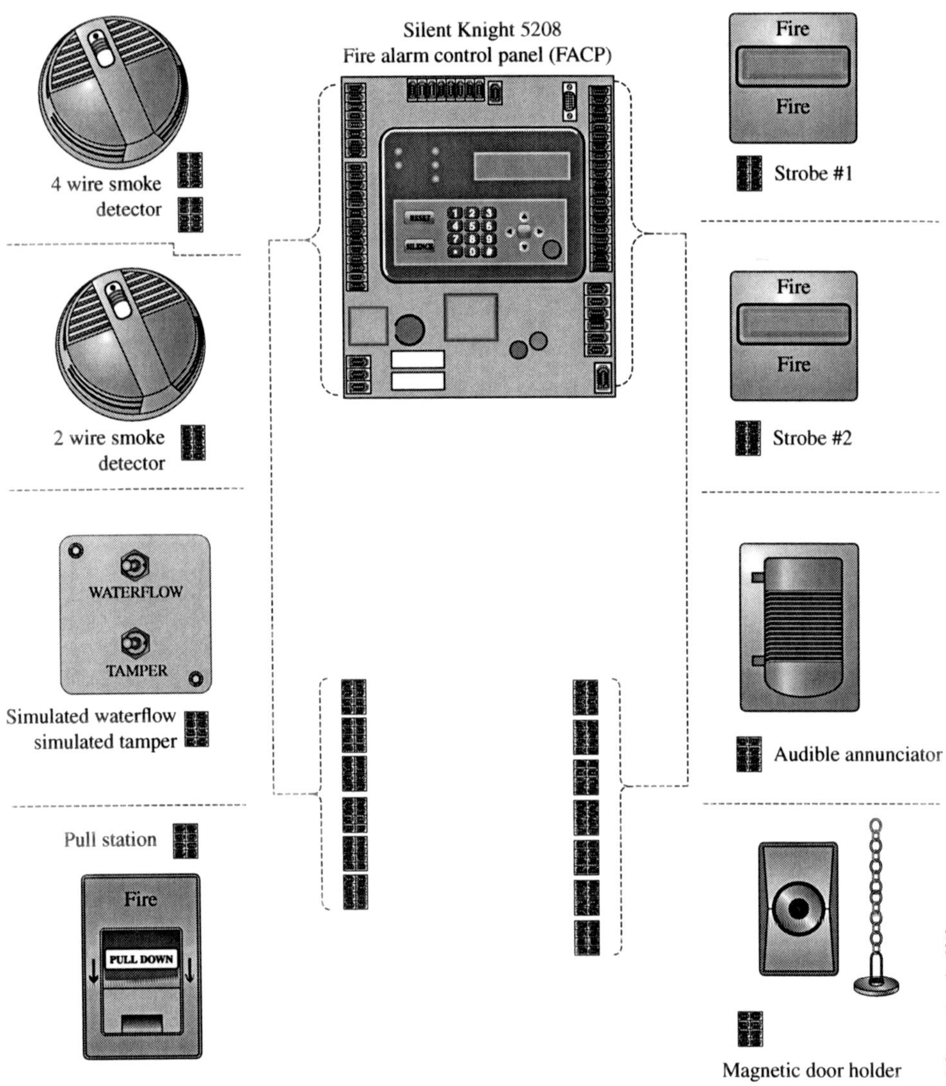

FIGURE 13–1 Fire alarm lab station

In addition, system devices are also mounted on the plywood, to each side of the control panel, and in like manner wired to external terminal strips. To complete the upcoming lab procedure, students wire between the external device terminal strips and the external control panel terminal strips in order to make the necessary connections.

A main ac power switch and a dc battery power switch are also needed so as to be able to connect and disconnect system power as required.

PRE-JOB QUESTIONS

1. Where would you find the ratings for a power-limited fire alarm power supply in the *NEC*?
2. Give examples of some fire alarm initiation devices.
3. Give examples of some fire alarm notification appliances.
4. What other types of circuits are controlled by the fire alarm output?
5. What is the difference between a Class-A (Style D) wired input device and a Class-B (Style B) wired input device?

OBJECTIVES

- Connect and test input and output devices of a fire alarm control panel (FACP).
- Configure inputs as either Class A (Style D) or Class B (Style B).
- Determine the differences among normal, trouble, supervisory alarm, and general alarm.
- Wire and test a PSR circuit.

REQUIRED MATERIALS

Silent Knight 5208 fire alarm panel

2-wire smoke detector—ESL 429CT

4-wire smoke detector—ESL 449CT

Simulated tamper input device—SPST toggle switch

Simulated waterflow input device—SPST toggle switch

Pull station device—FED 4050-001T-201

2 strobes—WHE RSS24MCW-FR

1 audible horn—System Sensor SpectrAlert H12/24 horn

1 magnetic door holder—EDW 1508-AQN5

12 4.7-KΩ resistors to be used for end-of-line (EOL) connections (a different value may be needed depending on type of system being used for this experiment)

Magnet to test smoke detectors
DPDT 24 VDC power supervision relay (PSR)
18 AWG, 2-conductor FPL cable
Screwdrivers
Wire cutter/stripper
Safety glasses
18 AWG terminal connector spades
Terminal connector crimp tool

TEST EQUIPMENT

Digital multimeter (DMM)

PROCEDURE

1. Ensure that the main AC panel power and DC battery backup power switches are both OFF.
2. Turn ON the main AC power switch to the panel.
   a. Examine the panel to identify any alarm or trouble conditions.
3. Turn ON the battery switch to enable the DC backup power.
   a. Examine the panel to identify any alarm or trouble conditions.
4. Wire the waterflow as Class A (Style D) to Zone 1, as shown in Figure 13–2. Check for continuity and shorts with the DMM.
5. Wire the tamper switch as Class A (Style D) to Zone 2, as shown in Figure 13–2. Check for continuity and shorts with the DMM.

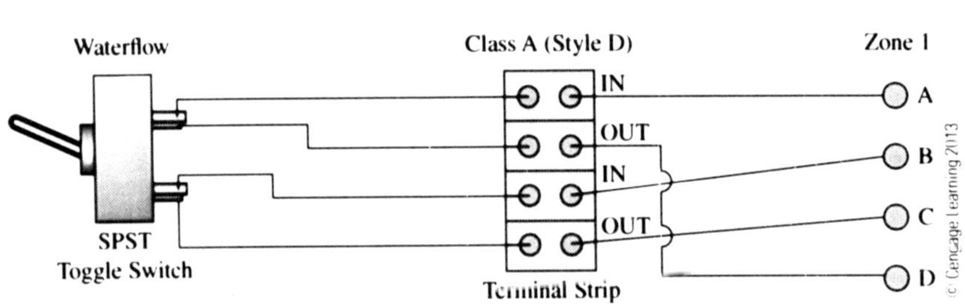

FIGURE 13–2 Waterflow and Tamper Connections

6. Wire the 4-wire smoke detector to power and Zone 3, as indicated in Figure 13–3, making sure to supervise power through an end-of-line relay. When wired correctly, you should be able to measure the EOL resistor before connecting to the fire alarm zone.

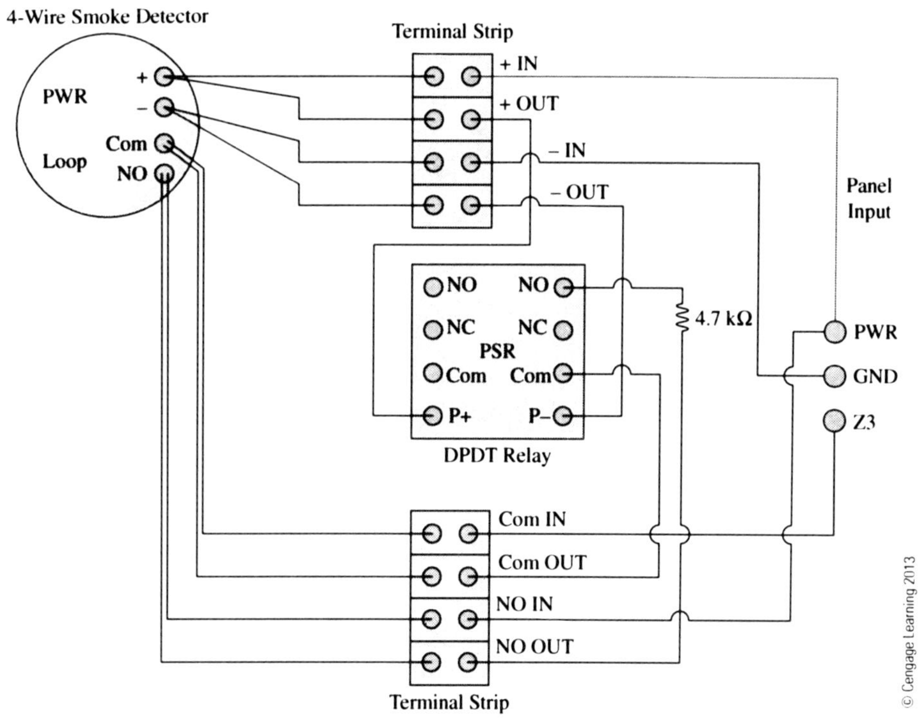

FIGURE 13–3 4-wire Smoke Detector with PSR

7. Wire the 2-wire smoke detector to Zone 4, as indicated by Figure 13–4. Make sure to measure the EOL resistor before connecting to the panel.

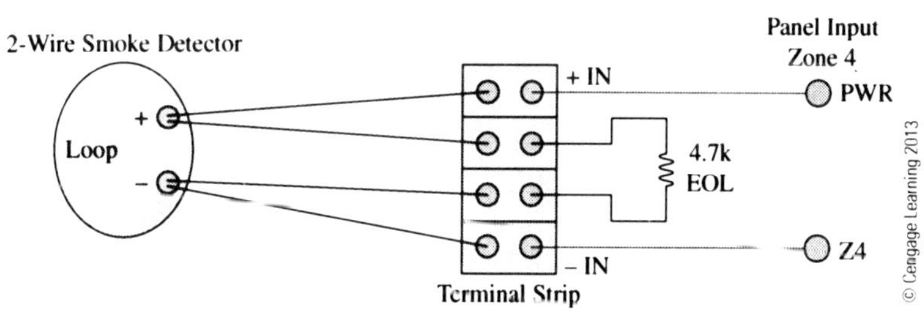

FIGURE 13–4 2-Wire Smoke Detector with EOL Resistor

8. Wire the pull station to Zone 5, as indicated by Figure 13–5. Make sure to measure the EOL resistor before connecting to the panel.

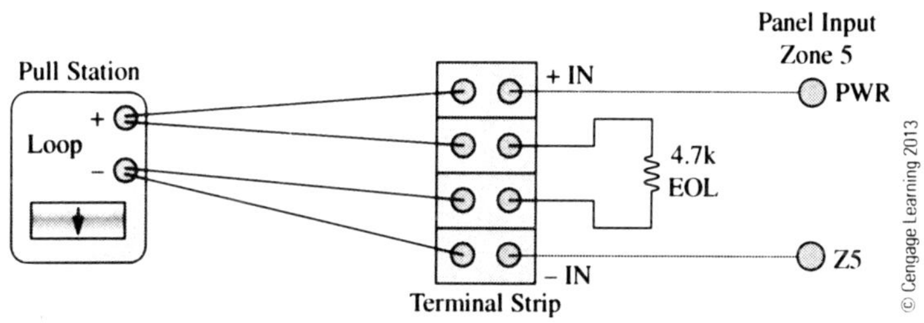

FIGURE 13–5

9. Install EOL resistors to unused Zones 6 through 10. Measure voltage before and after connecting EOL. Notice that the panel zones clear as EOL connections are added.
10. Install the EOL resistor to the top strobe and connect to NAC 1, as indicated by Figure 13–6. Measure EOL before connecting to the panel.

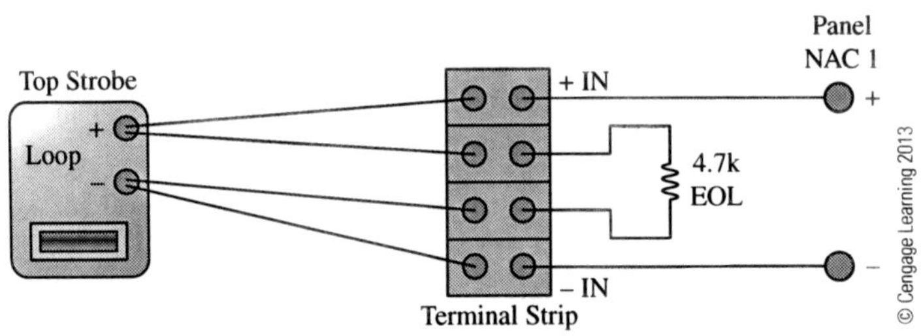

FIGURE 13–6 Top Strobe with EOL Resistor

11. Install EOL resistor on the horn, then connect the horn to the middle strobe, and then the strobe to NAC 2, as indicated by Figure 13–7.

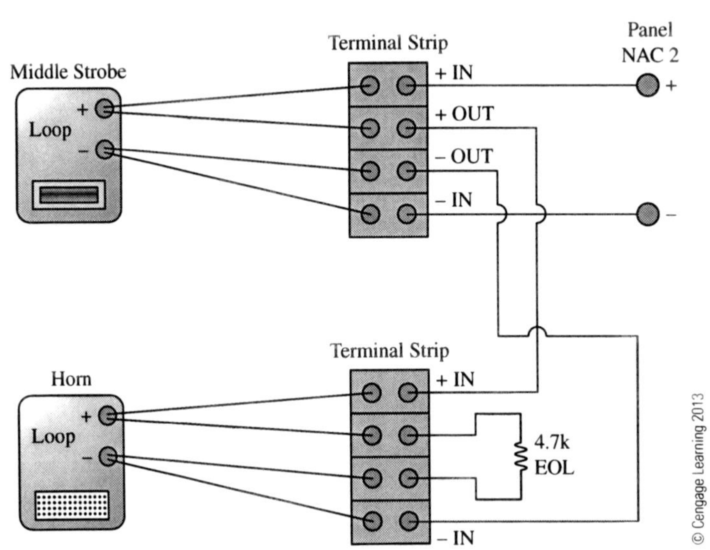

FIGURE 13–7 Middle Strobe with Horn and EOL Resistor

12. Install EOL resistors on unused NAC circuits 3 and 4. Measure the voltage before and after connecting EOL to the NAC circuit. Notice that panel troubles should clear and show normal.
13. Connect door holder to N/C contacts 43 and 44, and to 24 VDC panel power 27 and 26, as indicated by Figure 13–8. Remember to switch the positive side. Hook the door holder together. If wired correctly, the magnet will be energized and hold the metal plate.

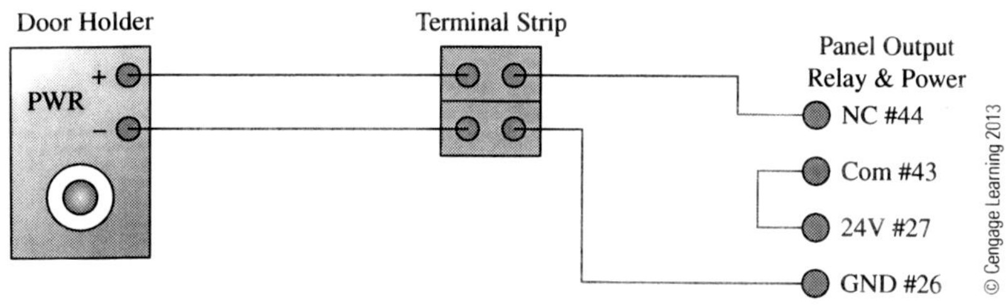

FIGURE 13–8 Door Holder to Panel Output Relay

14. Use a magnet to put the 4-wire smoke detector into the alarm.
15. Measure the voltage of the horn, strobe, and door holder.
    Horn __________ Strobe __________ Door holder __________
16. Measure the voltage across the 4-wire smoke detector EOL resistor.
    4-wire smoke detector EOL resistor __________ V
17. Press the RESET button to reset the panel. You will need to enter the code 1111.
18. Reconnect the door holder.
19. Remeasure the voltage of the horn, strobe, and door holder.
    Horn __________ Strobe __________ Door holder __________
20. Remeasure the voltage across the 4-wire smoke detector EOL resistor.
    4-wire smoke detector EOL resistor __________ V
21. Disconnect power to the power supervision relay (PSR).
22. What happens?
23. Press the ACK/Enter button to silence the panel.
24. Reconnect the power to the PSR.
25. Use a magnet to put the 2-wire smoke detector into alarm.
26. Verify that all circuits are working properly, and then press the RESET button to reset the panel. You will need to enter the code 1111.
27. Pull the pull station.
28. Press the RESET button to reset the panel. Enter the code 1111.
29. What happens?
30. Use a small Phillips screwdriver to reset the pull station. Press the RESET button on the panel after the pull station has been reset. Remember to use the code 1111.
31. Flip the tamper switch. Notice anything different?
32. Press the ACK/Enter button to silence the panel.
33. Reset the tamper switch. What happens?
34. Flip the waterflow switch. Press the ACK/Enter button to silence the panel.
35. Hit the SILENCE button, remembering to use the code 1111.
36. Notice anything different?
37. Reset the waterflow switch. Hit the SILENCE button and enter the code 1111.

38. Hit the RESET button, enter the code 1111, and reconnect the door holder.
39. Turn the battery backup power switch OFF.
40. Turn the main panel power switch OFF.
41. Remove all temporary jumpers that you have installed, and clean up the workstations.

FINAL SUMMARY

Waterflow and Tamper

In this experiment, waterflow and tamper are simulated with a toggle switch. For waterflow, the panel monitors the status of the input loop circuit through a Class A (Style D) connection. With the toggle switch in the open position, the system indicates a normal condition. When the toggle switch is moved from open to closed, it places a direct short on the input circuit, which indicates that water is now flowing somewhere in the building. Based on system programming, such a condition sends the system into general alarm.

A tamper switch is used to monitor the position of the water control valve to determine whether the valve is open or closed. A closed valve indicates that water has been shut off to the sprinkler heads in the building. Such a condition would be a potential hazard in the event of an actual fire. When the water valve is closed, the tamper switch changes state, and a short is placed on the input circuit. Based on system programming, such a condition forces the panel into a supervisory alarm, indicating that the water valve requires attention.

Four-Wire versus 2-Wire Smoke Detectors

Four-wire smoke detectors use two wires for power and two wires for the supervised loop circuit. Two-wire smoke detectors are loop powered and do not require a separate power connection. When 4-wire smoke detectors are connected as Class B (Style B) circuits, the power to the device must be supervised through a listed power supervision relay (PSR). A loss of power causes the PSR to change state, and the EOL resistor is disconnected from the input loop circuit. When this happens, the panel goes into trouble alert, indicating that there is a problem with the device. A 2-wire device is loop powered and does not require a PSR. The Class B (Style B) connection requires only that an EOL resistor be placed on the last device in the zone. Resistor values vary depending on manufacturer.

Pull Stations

Pull stations are simply shorting circuits, and they are wired as initiating input devices. The EOL resistor is used to supervise the connection to the input circuit. Because the pull station is in a normally open state, the EOL resistor provides the return path for supervision. The system goes into alarm when the handle on the pull station is pulled, and the device shorts the input circuit.

Once pulled, a pull station must be reset by using a screwdriver to press the reset button inside the device.

Horn/Strobes

Horn/strobes are notification appliance circuits (NAC), and they are connected to the output of the fire alarm control panel. A NAC circuit differs from an input circuit in that it uses polarity reversal to operate. A diode inside the horn/strobe prevents the device from turning on when the polarity of the output is in the normal, or non-alarm, state. The output terminals on the control panel are marked to identify the polarity of an alarm condition. When the control panel goes into alarm, the polarity of the output reverses. The diode is then forward biased and the device turns on. The EOL resistor provides a return path for current flow so that the panel can supervise and monitor the status of the output circuit.

Door Holders

Door holders are connected to the normally closed (NC) relay contacts at the output of the fire alarm panel. Power is looped through the relay circuit to energize the electromagnets within the door holders. When the panel goes into alarm, the output relay changes state and power is removed from the door holders so that the doors can close. Doors must be put back into holding position after the panel has been reset.

FINAL QUESTIONS

1. What happened to the voltage measurements of the horn and strobes after the panel went into alarm?
2. What happened when the power lead was disconnected from the PSR?
3. What is the purpose of adding a relay to the power leads of a 4-wire smoke detector?
4. At the panel, how would you identify which input or output zone is in trouble without looking at the panel display?
5. Why did we add end-of-line resistors to the input and output zones that are not being used?

Name ______________________

Date ______________________

Class ______________________

LAB 14

Anaerobic Fiber-Optic Connectors

REQUIRED READING

Introduction to Low Voltage Systems, 2nd edition, Chapter 10

INTRODUCTION

Before doing this lab, it is important to review a few safety precautions:

1. When working with fiber-optic cables, always wear safety glasses.
2. Each workstation should have its own fiber-optic trash bin. Always dispose of fiber debris properly.
3. Never touch your eyes when working with fiber, and when you are done, be sure to wash your hands.
4. Individuals wearing contact lenses should never remove or clean their lenses until they are absolutely sure that their hands are clean and free of any stray fiber particles.
5. Never look directly into the end of a fiber-optic strand until you are positive that it is disconnected from a light source. To do so could risk potential damage to your eye or possible blindness.
6. Not all light sources are visible. Because infrared light is invisible, it could damage your eye before you even realize that it has occurred. And in some cases, a light source may be supplied from a high-powered laser.
7. Always be sure the light source is removed first before installing, servicing, or repairing any system.
8. To prevent the accidental ingestion of glass particles, and possible contamination of cables, no food or beverage is allowed in the work area.
9. Wash your face and hands before working with fiber-optic cables and connectors.

During the connection process, the exposed fiber, buffer, or jacket should be cleaned repeatedly to remove any dust or oils that may have accumulated on the surface area. In addition, the MSDS sheets for Loctite® adhesive and primer should be available and read by the student before performing the lab exercise.

The following lab exercise outlines how to attach anaerobic connectors to fiber-optic cables. The anaerobic connectors used are for type 62.5/125 MM fiber-optic cable. Since there are differences between types FC, LC, SC, and ST connectors, the student should become familiar with the various parts included within the connector kit.

Each kit should include the following: large and small strain relief boot, crimp sleeve, connector body, and dust cover. The smaller strain reliefs are used on buffered fibers, and the larger are used on jacketed fibers. In addition, for types SC and FC connectors, an additional spacer is provided that is required on jacketed-type cables. And, for type SC only, an exterior housing attaches over the connector body during the final assembly.

The fiber-optic cables used in this experiment can be jacketed, such as a simplex patch cable, or unjacketed, as in the subgroup of buffered fibers from a breakout cable. Jackets have a 3-mm thickness and buffers, 900 μm. Jacket strippers are different from buffer strippers. To prevent damage to the fiber, they should not be used interchangeably.

Before beginning the lab, lay out all the tools and parts on a fiber safety mat, and be sure to have a fiber trash bin available at each workstation. Always remember to wear safety glasses.

PRE-JOB QUESTIONS

1. Name the parts of a fiber connector.
2. What does 62.5/125 refer to with regard to multimode fiber?
3. Describe the difference between an anaerobic connection and a Hotmelt® connection.
4. What type of epoxy is used for an anaerobic connector?
5. List three safety rules to follow when working with fiber-optic cable.

OBJECTIVES

- Follow all safety rules, and demonstrate safe working procedures.
- Identify the tools and materials needed for terminating a fiber-optic connector.
- Demonstrate the procedure for stripping, preparing, and attaching an anaerobic connector to a fiber-optic cable.
- Demonstrate the procedure for polishing the end of a fiber-optic connector.

REQUIRED MATERIALS

Safety glasses
Loctite 648 adhesive
Loctite 7649 primer
Loctite MSDS sheets
Adhesive bottle needles or adhesive syringe
Fiber safety mat
Fiber trash bin
Kevlar shears
Fiber tweezers
Fiber jacket stripper
Fiber buffer stripper
Fiber crimper
Fiber scribe
Multimode, 62.5/125 polishing puck
For LC connectors, Leviton® 1.25-mm polishing puck
Polishing pad
62.5/125, MM fiber-optic cable, with or without a 3-mm jacket
2 each LC, SC, FC, or ST Leviton Fast Cure Connectors
99 percent alcohol wipes
Lint-free wipes
0.3-μm film (blue)
3-μm film (yellow)
12-μm film (dark pink)

TEST EQUIPMENT

Fiber-optic light source and power meter, with types LC, SC, FC, or ST adaptors
Fiber-optic microscope (200×)

PROCEDURE

Prepare the Fiber

1. Gather all the tools you will need, and arrange them for easy use. It helps to be working in an area that has good light and is not dusty.
2. Cut a 3-foot section of 62.5/125 multimode fiber.

   SAFETY WARNING: *Working with fiber can be a hazard because of the small, sharp shards of glass that are produced when cutting or cleaving* *the strands. Always wear safety glasses to protect your eyes, and be careful not to get pieces of fiber stuck in your fingers or hands. If you do get a fiber sliver, use the tweezers to remove it. Always keep a disposable fiber trash bin nearby for all fiber scraps.*

3. Each connector comes in four parts: strain relief boot, crimp sleeve, connector body, and dust cap. Lay out all four parts of your connector. If you are using an SC connector, there is an additional exterior housing that slides over the connector body during the final assembly. Also, type SC and FC connectors have an extra part, an internal spacer that is used on jacketed fiber.
4. For 3-mm jacket cable, use the ribbed boot that has been provided with the connector. For 900-µm buffered cable, use the smooth strain relief boot. Slide the strain relief boot onto the fiber, narrow end first.
5. For jacketed fibers only, as shown in Figure 14–1, slide the crimp sleeve onto the fiber, narrow end first. The crimp sleeve is not used on unjacketed cables.

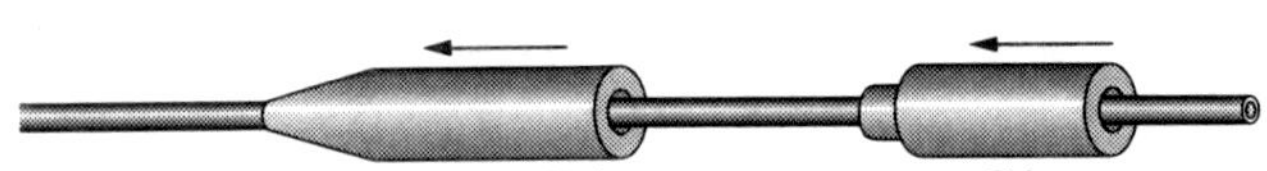

FIGURE 14–1 Jacketed Fiber Assembly

6. Slide the boot and crimp sleeve back along the fiber, out of the way, as you prepare to strip the jacket and buffer.
7. To avoid contamination, clean the cable jacket or exposed buffer with a 99 percent alcohol wipe, and then dry it with a lint-free wipe.
8. For jacketed fiber, mark the length of jacket to be removed, based on Figure 14–2.
9. Set the jacket cutting tool on the desired mark (#4 for 3-mm fiber), close down on the fiber at the desired location, and hold the tool closed. Do *not* try to strip the jacket off as you might with copper; you could damage the fiber.
10. Release the tool, and grab the jacket with your fingers. Gently twist and pull it from the fiber. It will slide off easily.
11. Trim the aramid strands with the Kevlar shears to a length of ¼ inch.
12. Hold the buffer stripper at a 45 degree angle to the fiber so that it does not bend the fiber when you clamp down.

    WARNING: *It is very important that care be taken so that the glass fiber does not break. Practice first on a short piece of fiber. Attempt to strip no more than 1/8 inch of buffer at a time until you get the feel of it; then you may be able to strip longer lengths without breaking the fiber.*

13. Squeeze the stripper firmly around the fiber to cut the buffer fully. Pull slowly and steadily, keeping the fiber straight as you strip off the buffer.

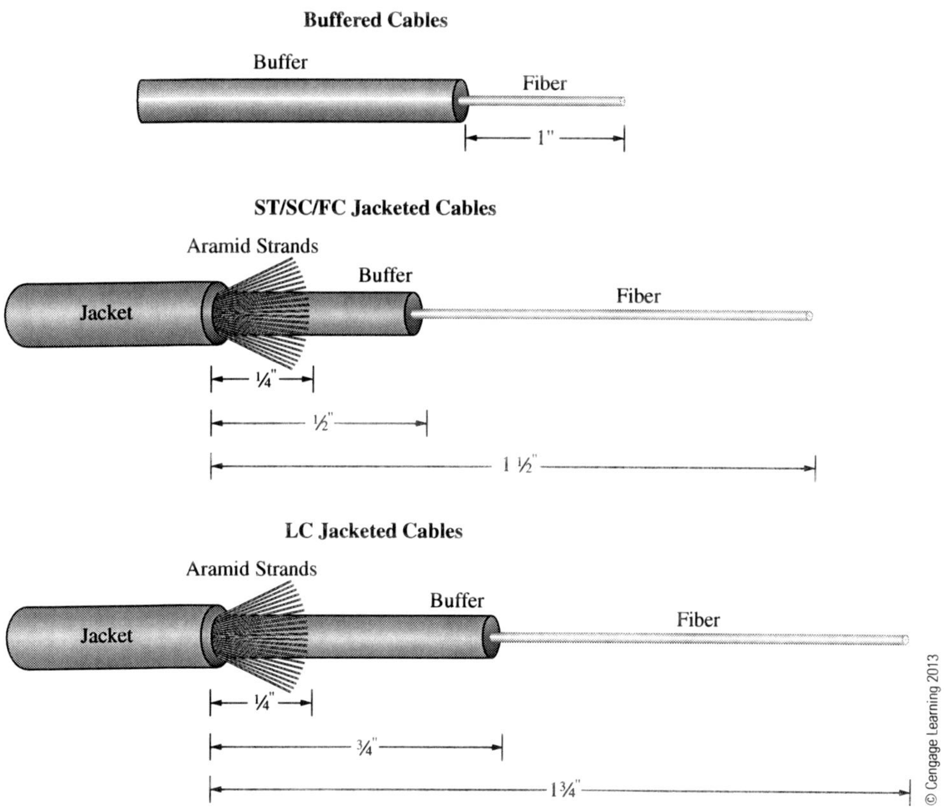

FIGURE 14–2 Measurements for stripped cables

It will *not* slide off easily like the jacket. Continue to take ⅛-inch to ¼-inch bites of buffer (3–6 mm). There must be enough exposed fiber for the connector body to adhere to. Remove the buffer until 1 inch of fiber is exposed, as shown in Figure 14–2.

14. Clean the jacket and exposed buffer with a 99 percent alcohol wipe, and then dry it off with a lint-free wipe.
15. For SC or FC type connectors, slide the spacer over the buffer, narrow side first, and push it all the way down until it rests against the jacket, as shown in Figure 14–3.

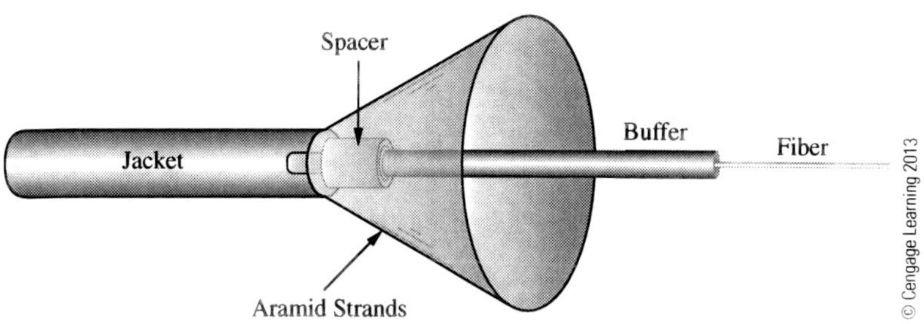

FIGURE 14–3 Type SC and FC connector spacer

Attach the Connector

16. Remove the dust cap from the connector.
17. Shake the Loctite 648 adhesive bottle before using. Remove the cap, and attach a needle to the top of the bottle. As an alternative, syringes can also be used.
18. Insert the needle into the back of the connector. Inject the adhesive into the connector, being careful not to overfill. Continue to fill until a dot of adhesive is visible on the other end.
19. Remove the needle, and wipe away excessive adhesive from the end of the ferrule.
20. Dip the cleaned fiber into the accelerator, or primer solution. Be sure to coat it completely, right up to the buffer.
21. Insert the fiber into the back of the connector.

    ***WARNING**: With this method, you must work fast and make sure the fiber is inserted rapidly or the adhesive will set before the fiber is fully inserted.*

22. Hold the fiber in place for 30 seconds to allow the fiber to bond with the connector.
23. For unjacketed fiber, slide the strain relief boot up over the back of the connector.
24. For jacketed fiber, slide the crimp tube over the aramid strands and connector.
25. Crimp the large end of the crimp tube with the crimp tool, using the 0.178-inch hex die, and then crimp the small end with the 0.151-inch hex die. For type LC connectors, use the 0.128-inch hex die.
26. Slide the strain relief boot over the back of the connector. For types LC and SC connectors, first ensure proper orientation of the body, and then slide the strain relief boot over the back of the connector.

Cleave and Polish

27. When the adhesive is dry, using the scribe tool, scribe the fiber where it meets the adhesive on the end of the ferrule.
28. Remove the scribed piece of fiber from the connector, and dispose of it in the fiber trash bin.
29. Air polish the tip of the connector with a 12-μm film. The 12-μm film is dark pink in color. Hold the film with the thumb and forefinger, and gently rotate around the end of the connector for 10–20 seconds, making 18–20 1-inch circles.
30. Gently press the end of the connector into a lint-free wipe. Twist in place to remove any debris. Do not drag the connector across the wipe.
31. Wipe the bottom surface of the polishing puck and the surface of the polishing pad with a 99 percent alcohol wipe. For LC connectors, use Leviton's 1.25-mm polishing puck.

32. Place the 3-μm film (yellow) on the polishing pad, dull side up, and then place the polishing puck on top of the film.
33. Gently place the connector into the puck.
34. Polish the end of the connector, using very light pressure. Trace 15–20 figure-eights over the film. Continue to polish until the fiber is flush with the end of the connector.
35. Check the end of the connector with the fiber microscope. It should look smooth. Compare the image of a polished connector with an unpolished connector.
36. Repeat the polishing steps with the blue 0.3-μm film.
37. Connect the cable to a light source and power meter to verify the connector.

FINAL SUMMARY

Quick curing adhesives can vary. Some cure too fast for the novice, who often finds the fiber stuck halfway into the connector when assembling. It is also more difficult to achieve a hard enough bead on the end of the ferrule, making these adhesives more difficult to cleave and polish. However, once you master them, they can provide the same low loss output as epoxy connectors.

Always remember to work in a clean environment and to wear safety glasses at all times. Small pieces of glass fiber can get under your skin or in your eyes and cause irritation and possible damage. When you are finished, be sure to clean up the work area, and dispose of all waste appropriately. And be sure to wash your hands when finished.

FINAL QUESTIONS

1. The procedure just outlined can be broken down into three parts. What are they?
2. How does preparing a jacketed cable differ from preparing a nonjacketed cable?
3. Besides anaerobic, what other types of adhesives are used on fiber-optic connectors?
4. What should you do if the connector fails the light source power meter test?
5. Which part of the process did you find the most challenging to perform?

Name ______________________

Date ______________________

Class ______________________

LAB 15

Hot Melt® Fiber-Optic Connectors

REQUIRED READING

Introduction to Low Voltage Systems, 2nd edition, Chapter 10

INTRODUCTION

Before doing this lab, it is important to review a few safety precautions:

1. When working with fiber-optic cables, always wear safety glasses.
2. Each workstation should have its own fiber-optic trash bin. Always dispose of fiber debris properly.
3. Never touch your eyes when working with fiber, and when you are done, be sure to wash your hands.
4. Individuals wearing contact lenses should never remove or clean their lenses until they are absolutely sure that their hands are clean and free of any stray fiber particles.
5. Never look directly into the end of a fiber-optic strand until you are positive that it is disconnected from a light source. To do so could risk potential damage to your eye or possible blindness.
6. Not all light sources are visible. Because infrared light is invisible, it could damage your eye before you even realize that it has occurred. And in some cases, a light source may be supplied from a high-powered laser.
7. Always be sure the light source is removed first before installing, servicing, or repairing any system.
8. To prevent the accidental ingestion of glass particles, and possible contamination of cables, no food or beverage is allowed in the work area.
9. Wash your face and hands before working with fiber-optic cables and connectors.

During the connection process, the exposed fiber, buffer, or jacket should be cleaned repeatedly to remove any dust or oils that may have accumulated on the surface area.

The following lab exercise will outline how to attach a Hot Melt connector to a fiber-optic cable. A Hot Melt connector differs from an anaerobic or heat-cured epoxy connector in that the connector is prefilled with an adhesive. A Hot Melt oven is then used to soften the adhesive before inserting the fiber into the connector.

A special note about curing ovens: A Hot Melt oven is different from a heat-cured epoxy oven. Oven temperatures differ between the two types, and they should not be used interchangeably. A heat-cured epoxy oven will not get hot enough to soften the adhesive in a Hot Melt connector, and a Hot Melt oven will get too hot for a heat-cured type epoxy, and the epoxy will be destroyed. For this lab exercise, make sure that you are using a Hot Melt oven, and that the connectors are prefilled Hot Melt connectors.

The connectors used in this lab are type ST Hot Melt connectors for 62.5/125 MM fiber-optic cables. Each connector kit should include the following: a strain relief boot, a ¾-inch-long clear strain relief tube, connector body, and dust cover. The strain relief tube is not used on jacketed fibers. Compared to the anaerobic connector kits, the Hot Melt connector kits do not include a crimp sleeve, and therefore a crimp tool will not be required.

Before beginning the lab, lay out all the tools and parts on a fiber safety mat, and be sure to have a fiber trash bin available at each workstation. And remember to always wear safety glasses.

The fiber optic can be jacketed, such as in a simplex patch cable, or unjacketed, as in the subgroup of buffered fibers from a breakout cable. Jackets have a 3-mm thickness, and buffers, 900 μm. Jacket strippers are different from buffer strippers. To prevent damage to the fiber, they should not be used interchangeably.

PRE-JOB QUESTIONS

1. Name the parts of the fiber connector that were included with the Hot Melt connector kit.
2. Can a multimode (MM) fiber connector be attached to a strand of single-mode (SM) fiber?
3. How does the Hot Melt connector kit compare with the anaerobic connector kit?
4. Each Hot Melt connector is prefilled with epoxy. How do you make the epoxy soft enough to be able to insert the fiber?
5. List three safety rules when working with fiber-optic cable.

OBJECTIVES

- Follow all safety rules and demonstrate safe working procedures.
- Identify the tools and materials needed for terminating a fiber-optic connector.

- Demonstrate the procedure for stripping, preparing, and attaching a Hot Melt connector to a fiber-optic cable.
- Demonstrate the procedure for polishing the end of a fiber-optic connector.

REQUIRED MATERIALS

Safety glasses
Fiber safety mat
Fiber trash bin
Kevlar shears
Fiber tweezers
Fiber jacket stripper
Fiber buffer stripper
Fiber scribe
Multimode, 62.5/125 polishing puck
Polishing pad
62.5/125 MM fiber-optic cable, with or without a 3-mm jacket
2 each ST Hot Melt Connectors
Hot Melt oven with type ST connector holder and cooling rack
99 percent alcohol wipes
Lint-free wipes
0.3-μm film (blue)
3-μm film (yellow)
12-μm film (dark pink)

TEST EQUIPMENT

Fiber-optic light source and power meter with type ST adaptors
Fiber-optic microscope (200×)

PROCEDURE

Prepare the Fiber

1. Gather all the tools you will need, and arrange them for easy use. It helps to be working in an area that has good light and is not dusty.
2. Cut a 3-foot section of 62.5/125 multimode fiber.

SAFETY WARNING: *Working with fiber can be a hazard, due to the small, sharp shards of glass that are produced when cutting or cleaving the strands! Always wear safety glasses to protect your eyes and be careful to not get pieces of fiber stuck in your fingers or hands. If you do receive a fiber sliver, use the tweezers to remove it. Always keep a disposable fiber trash bin nearby for all fiber scraps.*

3. Each connector comes in four parts: strain relief boot, clear strain relief tube, connector body, and dust cap. Lay out all four parts of your connector.
4. Slide the strain relief boot onto the fiber, narrow end first.
5. For unjacketed, buffered fiber only, slide the strain relief tube onto the fiber. The strain relief tube is not used on jacketed fiber.
6. Slide the strain relief boot and tube back along the fiber, out of the way, as you prepare to strip the jacket and buffer.
7. To avoid contamination, clean the cable jacket or exposed buffer with a 99 percent alcohol wipe and then dry it with a lint-free wipe.

Jacketed Fiber

8. Mark the fiber jacket to be removed ¾ inch, based on Figure 15–1.

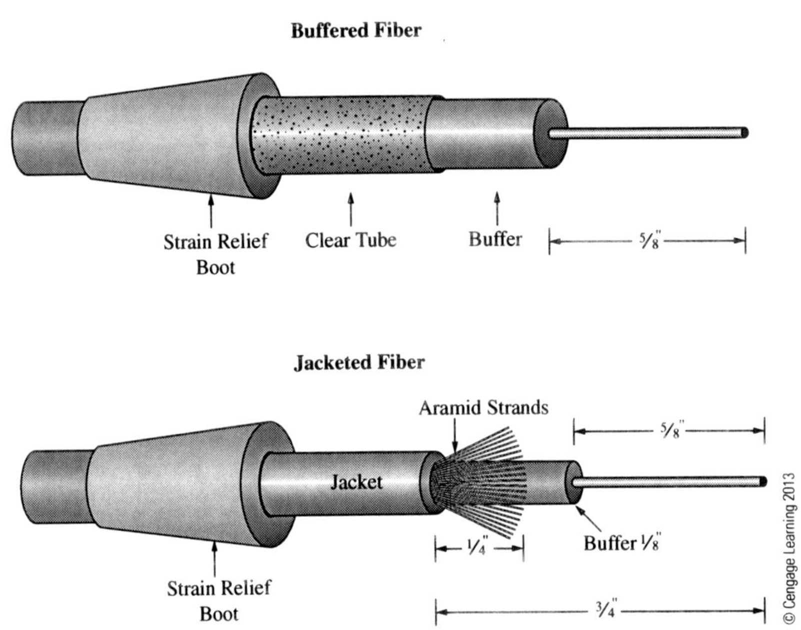

FIGURE 15–1 Measurements for stripped cables when using Hot Melt connectors.

Set the jacket cutting tool on the desired mark (#4 for 3-mm fiber), close down on the fiber at the desired location, and hold the tool closed. Do *not* try to strip the jacket off as you might with copper; you could damage the fiber.

9. Release the tool, and grab the jacket with your fingers. Gently twist and pull it from the fiber. It will slide off easily.
10. Trim the aramid strands with the Kevlar shears to a length of ¼ inch.
11. Hold the buffer stripper at a 45 degree angle to the fiber so that it does not bend the fiber when you clamp down.

WARNING: It is very important that care be taken so that the glass fiber does not break. Practice first on a short piece of fiber. Attempt to strip no more than 1/8 inch of buffer at a time until you get the feel of it; then you may be able to strip longer lengths without breaking the fiber.

12. Squeeze the stripper firmly around the fiber to cut the buffer fully. Pull slowly and steadily, keeping the fiber straight as you strip off the buffer. It will *not* slide off easily like the jacket. Continue to take 1/8-inch to 1/4-inch bites of the buffer (3–6 mm). Remove 5/8 inch of buffer to expose the fiber. One-eighth inch of buffer remains, as shown in Figure 15–1.
13. Clean the jacket and exposed buffer with a 99 percent alcohol wipe, and then dry it off with a lint-free wipe.

Buffered Fiber

14. Hold the buffer stripper at a 45 degree angle to the fiber so that it does not bend the fiber when you clamp down.

    WARNING: It is very important that care be taken so that the glass fiber does not break. Practice first on a short piece of fiber. Attempt to strip no more than 1/8 inch of buffer at a time until you get the feel of it; then you may be able to strip longer lengths without breaking the fiber.
15. Squeeze the stripper firmly around the fiber to cut the buffer fully. Pull slowly and steadily, keeping the fiber straight as you strip off the buffer. It will *not* slide off easily like the jacket. Continue to take 1/8-inch to 1/4-inch bites of the buffer (3–6 mm). Remove 5/8 inch of buffer to expose the fiber, as shown in Figure 15–1.

Attach the Connector

16. Remove the dust cap from the connector.
17. Preheat the Hot Melt oven.
18. Place the ST connector into the top of the connector holder, and insert it into the preheated oven. The connector is keyed; be sure to twist it into the bayonet until it latches.
19. Wait 3–5 minutes; the epoxy should be soft. Do not overheat or the epoxy will start to drip out the front of the connector. Never heat beyond 10 minutes.
20. Remove the connector holder from the oven. Use caution—it is very hot.

Unjacketed, Buffer Fiber

21. Gently insert the fiber into the back of the connector. Push it all the way up to the buffer until it stops.
22. Slide the clear strain relief tube up into the back of the connector until it stops.

Jacketed Fiber

23. Gently insert the fiber into the back of the connector. Push it all the way up until the aramid strands are inside the connector.
24. If you've done it correctly, the bare fiber should be visible on the bottom side of the connector holder.
25. Adjust the fiber into the vertical support connected to the handle of the connector holder. It will prevent the cable from bending past the allowed bend radius while the epoxy is cooling.
26. Carefully, move the connector holder to the cooling stand, and allow it to cool for 5 minutes.
27. After the connector has cooled, remover the connector from the connector holder, and slide the strain relief boot up over the back of the connector.

Cleave and Polish

28. When the epoxy is dry, using the scribe tool, scribe the fiber where it meets the end of the ferrule.
29. Remove the scribed piece of fiber from the connector, and dispose of it in the fiber trash bin.
30. Air polish the tip of the connector with a 12-μm film. The 12-μm film is dark pink in color. Hold the film with the thumb and forefinger, and gently rotate around the end of the connector for 10–20 seconds, making 18–20 one-inch circles.
31. Gently press the end of the connector into a lint-free wipe. Twist in place to remove any debris. Do not drag the connector across the wipe.
32. Wipe the bottom surface of the polishing puck and the surface of the polishing pad with a 99 percent alcohol wipe.
33. Place the 3-μm film (yellow) on the polishing pad, dull side up, and then place the polishing puck on top of the film.
34. Gently place the connector into the puck.
35. Polish the end of the connector, using very light pressure. Trace 15–20 figure-eights over the film. Continue to polish until the fiber is flush with the end of the connector.
36. Check the end of the connector with the fiber microscope. It should look smooth. Compare the image of a polished connector with an unpolished connector.
37. Repeat the polishing steps with the blue, 0.3-μm film.
38. Connect the cable to a light source and power meter to verify the connector.

FINAL SUMMARY

When working with Hot Melt connectors, it's all about timing—knowing how long to heat the connector before the fiber can be inserted and how long to cool them before polishing. Once you get a feel for the process, you will be able to attach multiple connectors in a reasonable amount of time.

Always remember to work in a clean environment and to wear safety glasses at all times. Small pieces of glass fiber can get under your skin or in your eyes and cause irritation and possible damage. When you are finished, be sure to clean up the work area, and dispose of all waste appropriately. And be sure to wash your hands when finished.

FINAL QUESTIONS

1. The above procedure can be broken down into three parts. What are they?
2. How does preparing a jacketed cable differ from preparing a nonjacketed cable?
3. How did you find the Hot Melt process compared with the anaerobic process with regard to level of difficulty and the amount of time it took to attach the connector?
4. Which part of the process did you find the most challenging to perform?

Name ______________________

Date ______________________

Class ______________________

LAB 16

Telephone Cables and Connectors

REQUIRED READING

Introduction to Low Voltage Systems, 2nd edition, Chapter 11

INTRODUCTION

Making and testing telephone cables is a common task for any telecommunications technician. For most residential and commercial installations, the cables extending from the wall outlet to the telephone equipment are typically flat multiwire cable. Such cables are terminated with type RJ-11 modular plugs. Depending on the number of telephone lines, the cable has either four or six conductors.

The RJ-11 modular plug is a 6-position plug that can have two, four, or six conductors running through it. When you look at a cable that has been preassembled, you can determine the number of conductors in the wire by counting the number of copper pins on the plug. When ordering the plugs, you refer to them as RJ-11 6P2C, 6P4C, or 6P6C.

Technically, the 4C plugs are RJ-14, and the 6C plugs are RJ-25. The following chart displays the pin-out and color codes for the various options.

Position	Pair	T/R	+ −	RJ-11	RJ-14	RJ-25	25-pair Color Code	US Bell Colors
1	3	T	+			T3	Wh/Gr	White
2	2	T	+		T2	T2	Wh/Or	Black
3	1	R	−	R1	R1	R1	Bl/Wh	Red
4	1	T	+	T1	T1	T1	Wh/Bl	Green
5	2	R	−		R2	R2	Or/Wh	Yellow
6	3	R	−			R3	Gr/Wh	Blue

When looking at the back of the plug where the cable enters, with the key tab pointed down, pin 1 is on the left. From the front of the plug, pin 1 is on the right. Figure 16–1 illustrates the pin order for an RJ-11 modular plug.

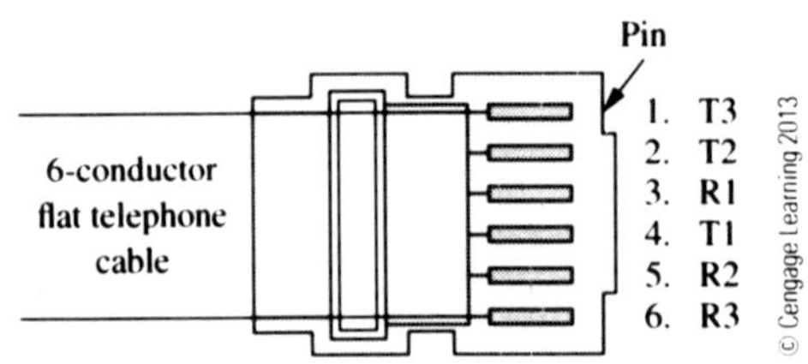

FIGURE 16–1

Connectors can be installed in one of two ways, either as straight through or cross-pinned. Straight through means that each pin is connected to the same pin number on each end of the cable. Cross-pinned means the order is reversed between each side of the cable. Cables that connect the telephone to the modular wall jack use the cross-pinned variety.

Modular Crimping and Stripping Tools

Some modular crimping tools have a cable cutter and stripper built into the tool, along with the crimp dies for size RJ-11 or RJ-45 modular plugs. Also, some crimping tools require you to change out the crimping die, depending on what size is required. Others have all three types, 6P, 4P, and 8P, built into the tool, and you simply need to choose the correct die position when using the tool.

If the cutter and stripper are built into your crimp tool, they will be located near the center of the tool. If you do not have a cutter on your tool, any standard wire cutter will work, provided you cut the end of the flat cable as straight as possible.

When stripping, locate the area on the crimping tool marked as stripper. Before inserting the cable, squeeze the tool handles together and notice that there is a thin gap remaining between the blades. The gap is calibrated to the thickness of the jacketing of the flat cable.

To strip the cable, insert the flat cable into the location of the tool marked "stripper." Push the cable all the way up until it stops inside the tool. The internal depth is set to strip off just the required amount of jacketing. Be sure that the flat sides of the cable are aligned against the cutting blades. Squeeze the handles of the crimp tool together. The blades will cut to the exact depth of the insulation, on each side of the flat cable. Release the tool handles, pull out the cable, and then remove the cut piece of jacket from the cable to expose the inner conductors.

PRE-JOB QUESTIONS

1. How many positions are there on an RJ-11 connector?
2. How many conductors are there on an RJ-14 connector?

3. What is the maximum number of phone lines that an RJ-11 modular plug can accommodate?
4. Look at the connectors on a phone cable for a computer modem. Determine what type of connector it is and how many lines are running in the cable. Also, is it straight through or cross-pinned?

OBJECTIVES

- Identify the differences between RJ-11, RJ-14, and RJ-25 modular plugs.
- Identify the difference between a straight through cable and a cross-pinned cable.
- Strip flat multiconductor cable.
- Attach modular plugs using an RJ-11 modular crimp tool.
- Test the cable for continuity and functionality with a cable tester.

REQUIRED MATERIALS

Computer modem phone cable—pre-assembled
3 feet of 4-conductor flat telephone cable
Wire cutters
RJ-11 modular plug stripping and crimping tool
Four RJ-11 (P6C4) connectors

TEST EQUIPMENT

Ideal, LinkMaster Pro XL cable tester

PROCEDURE

1. Using a wire cutter or the cutter included in the crimping tool, cut a 3-foot length of the 4-conductor flat cable.
2. Check both ends of the cable; be sure that they have been cut straight and even.
3. Insert one end of the flat cable into the stripper section of the crimping tool until it stops against the inside of the tool. Be sure that the flat sides of the cable are aligned against the cutting blades. Squeeze the handles of the crimp tool together. Release the tool handles, pull out the cable, and then remove the cut piece of jacket from the cable to expose the inner conductors.

4. Insert the stripped end of the flat cable into a RJ-11 connector. Be sure that the wires are securely pushed to the very front of the connector.
5. For 4-conductor cable, insert as following:

   Pos 1 = Empty
   Pos 2 = Black or Wh/Or
   Pos 3 = Red or Blu/Wh
   Pos 4 = Green or Wh/Blu
   Pos 5 = Yellow or Or/Wh
   Pos 6 = Empty
6. If your crimping tool has a removable die, check what type of die is inserted in the tool. Be sure the die is a type RJ-11, 4P. If not, then change out the die and insert the correct one.
7. Insert the RJ-11 modular plug into the 4P die of the crimping tool; it clicks into position.
8. Squeeze the tool tightly and then release until it ratchets back open.
9. Remove the connector from the crimping tool.
10. Modular plugs can only be crimped once. If you made a mistake, then cut off the connector, throw it away, and start over.
11. **Repeat steps 1–8** for the other side of the cable. However, this time insert the cable to accommodate the following color code for a cross-pinned cable:

    Pos 6 = Empty
    Pos 5 = Black or Wh/Or
    Pos 4 = Red or Blu/Wh
    Pos 3 = Green or Wh/Blu
    Pos 2 = Yellow or Or/Wh
    Pos 1 = Empty

Test the Cable

12. Connect one end of the completed cable to the RJ-11/RJ-45 adapter of the Ideal, LinkMaster Pro XL Cable Tester.
13. Insert the RJ-45 end of the adapter into the main port on the tester.
14. Connect the other end of the cable to the 6-position side of one of the remote terminators included with the tester kit. The tester includes 8 remote terminators that are used to load the end of the cable under test.
15. Leave the Remote 1 port empty.
16. Press the MODE button until you read Cable Test on the LCD display.
17. Press the select button SEL.
18. Read the display. It shows you the pin out at the near and far end of the cable, and also indicates whether the pins are reversed.
19. Does your cable pass the test for a cross-pinned cable? ____________
20. Press and hold the MODE button to shut off the tester, and press SEL.
21. Disconnect all wires, clean up your lab bench, and put away all tools and equipment.

FINAL SUMMARY

In the telecommunications industry, making test cables is a daily occurrence. It is not only important to know how to construct the cable, but it is also important to know how to use the proper tools. Once the test cable is complete, it is always important to test the cable with a cable tester to be sure you have made solid connections on each side and that the conductors are in the correct positions on each end of the cable.

FINAL QUESTIONS

1. How long did it take you to make your first cable?
2. Did you encounter any problems? If so, what were they, and do you feel confident that you can repeat the process next time without making a mistake?

Name ____________________

Date ____________________

Class ____________________

LAB 17

Telephone Local Loop Installation

REQUIRED READING

Introduction to Low Voltage Systems, 2nd edition, Chapter 11

INTRODUCTION

Telephone wiring comes in two varieties, residential and commercial. The traditional residential style wiring is comprised of a 4-wire line coming into the house from the telephone company. The line enters from the outside, connects to a primary protector, and then terminates onto a 42A block or RJ-11 modular outlet. The primary protector guards against any potential lightning strikes, and it can be placed on the inside or the outside of the house, just as long as it's as close as practicable to the point of entrance.

The 42A block or main RJ-11 outlet is regarded as the demarc. All wiring past the demarc is the responsibility of the homeowner, and all wiring up to the demarc is the responsibility of the telephone company. Figure 17–1 illustrates a typical residential installation.

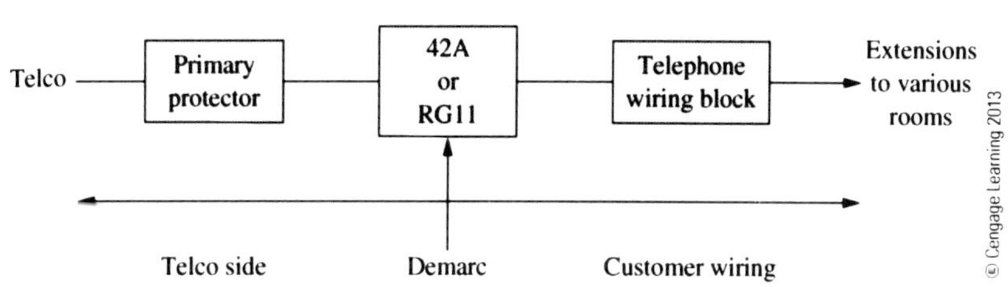

FIGURE 17–1

A commercial telephone installation is different. Most commercial businesses use multiple lines and connect the building through their own private PBX. The telephone company then provides a trunk line, which represents the main artery of phone lines entering the building from the outside. The size and number of trunk lines varies depending on the needs of the client and the size of the building.

The wiring for a typical commercial phone installation enters the building from the outside, terminates onto a splicing block, and then to a frame of primary protectors. From the primary protectors, the connections are then terminated onto type 66 connection blocks or binding posts, which represent the demarc. From the demarc, the connections are then routed to the main distribution frame (MDF) for distribution throughout the building structure. Figure 17–2 illustrates a typical commercial installation.

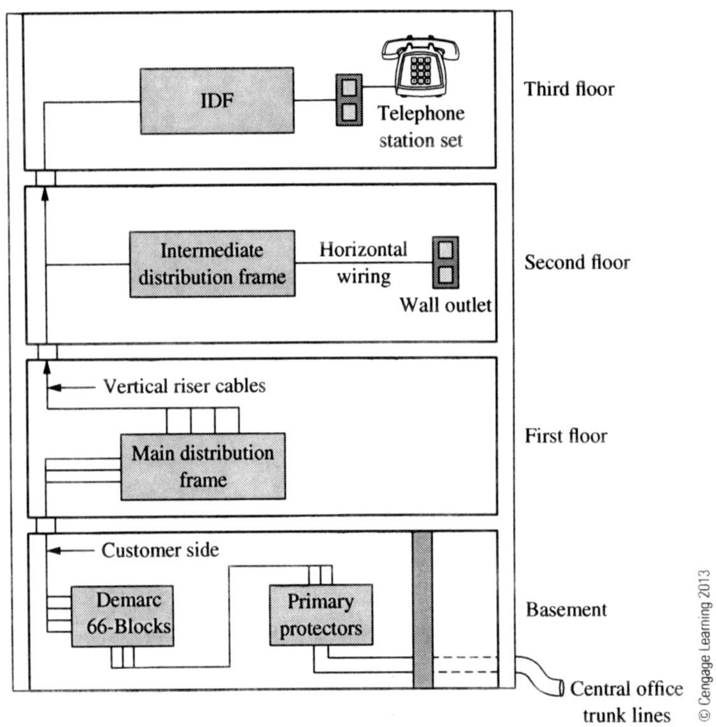

FIGURE 17–2

The benefit to using a type 66 connection block is that the individual phone wires do not need to be stripped, as they would if you were connecting to the screw terminals of a 42A block or binding posts. To make a connection, the insulated wire is placed over a lug on the 66 block, and it is punched down with an impact tool. The impact tool simultaneously pushes the wire into the connection lug and trims the excess from the end. Contact with the internal copper wire is made as the V-shaped lug of the 66 block cuts through the insulation of the wire, just far enough to make connection. A spudger tool, which looks like a pick, is then used if there are any stray pieces of wire left in the block that need to be removed, which could potentially short between lugs.

When you are using the impact punch tool, be sure that the cutting edge of the blade is positioned to the side of the wire to be trimmed. Notice that the cutting edge of the blade is sharp and extends past the opposite side of the blade. Figure 17–3 illustrates how to position the impact tool.

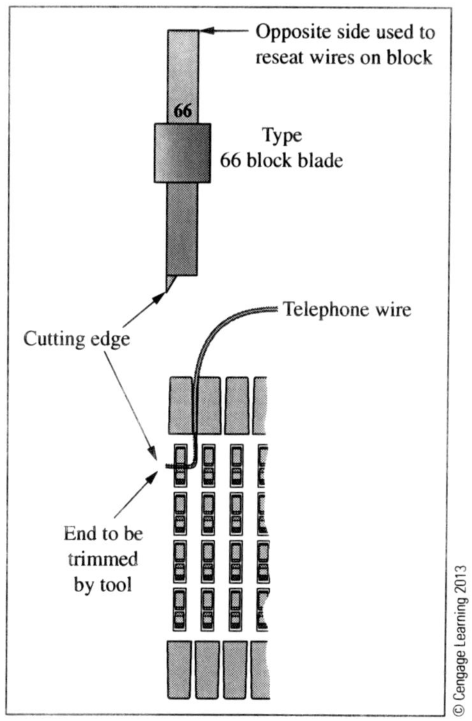

FIGURE 17–3

In this lab exercise, the student connects and tests a basic local loop. The Ring It telephone simulator is fully functional and represents the connection to the phone company. When all the connections are made, the test phones will be able to call each other.

Instructor Connections

Before doing the lab, the instructor is responsible for making permanent connections to the Ring It telephone simulator, the test phone and wall outlet, 66-block #1, and the input side of the 66-block primary protector. Be sure that the Ring It telephone simulator is set to mode "n," for "normal," to simulate a basic analog loop start phone line.

The ouput side of the 66-block primary protector represents the demarc.

Student Connections

The student connections are between the demarc and 66 block #2, and 66 block #2 and the station set wall outlet.

Figure 17–4 illustrates the connections.

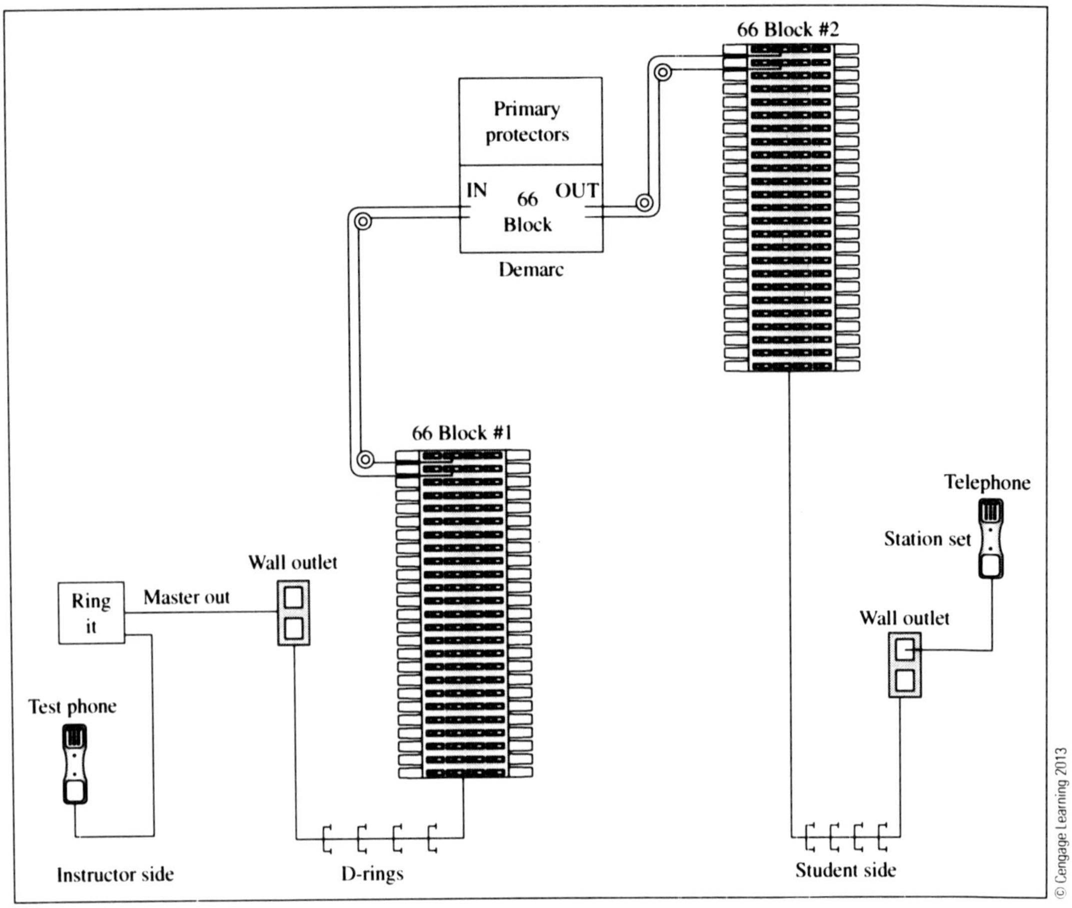

FIGURE 17–4

PRE-JOB QUESTIONS

1. What is meant by *local loop*?
2. Explain the term *trunk line*.
3. A PBX typically uses digital telephone lines between the PBX and station sets. Are digital lines compatible with analog lines?

OBJECTIVES

- Install a basic analog local loop.
- Install and test a primary protector.
- Install 6-conductor telephone cable onto a type 66 block.
- Install cross-connect wires between 66 blocks.
- Install RJ-11 wall outlets.
- Test the local loop with a telephone butt-set.

REQUIRED MATERIALS

Punch-down tool
Blades for type 66 connection blocks
1 roll of 6-conductor wire (red, green, yellow, black)
1 roll of 24 AWG, 2-pair cross-connect wire
2-Type 66 punch-down blocks and wall mounts
Type 66 punch-down lightning protection block
Two RJ-11 wall jacks
Wire stripper
Spudger
D-rings
Cable routing spools

TEST EQUIPMENT

Fluke TS22 test set (butt-set)
Ring It telephone simulator
Two telephone handsets
Digital multimeter (DMM)

PROCEDURE

Connect the Wall Outlet to the 66 Block

1. Cut a 3-foot section of 6-conductor telephone cable.
2. Using the wire stripper, strip about 2 inches of jacketing from one end of the cable.
3. Run the cable up through the bottom of 66 block #2 and out the upper left side of the mount.
4. Terminate the six conductors on to the first six rows along the left side. Use the following color code:

   Lug 1 = Green or White/Blue = T1
   Lug 2 = Red or Blue/White = R1
   Lug 3 = Black or White/Orange = T2
   Lug 4 = Yellow or Orange/White = R2
   Lug 5 = White or White/Green = T3
   Lug 6 = Blue or Green/White = R3
5. Rout the other end of the cable through the D-rings and over to the RJ-11 wall outlet mounted on the electrical box, just to the right.
6. Strip about 2 inches of jacketing from the end of the cable to expose the six conductors.

7. Using the wire strippers, strip each conductor about ½ inch, and connect to the screw terminals on the back side of the wall outlet. Use the following color codes:

   Pos 1 = White or White/Green
   Pos 2 = Black or White/Orange
   Pos 3 = Red or Blue/White
   Pos 4 = Green or White/Blue
   Pos 5 = Yellow or Orange/White
   Pos 6 = Blue or Green/White

8. When complete, screw the wall outlet down onto the box, and attach the cover to the wall outlet.

Locate Dial Tone

9. Locate the 66-block primary protector.
10. Figure 17–5 illustrates the configuration of the 66-block primary protector. It is different from a normal 66 block. The block is only 25 rows long, and the tip and the ring for each line share the same row. Two lugs are for input and two lugs are for output. The fuse for each line makes connection between the input and output.

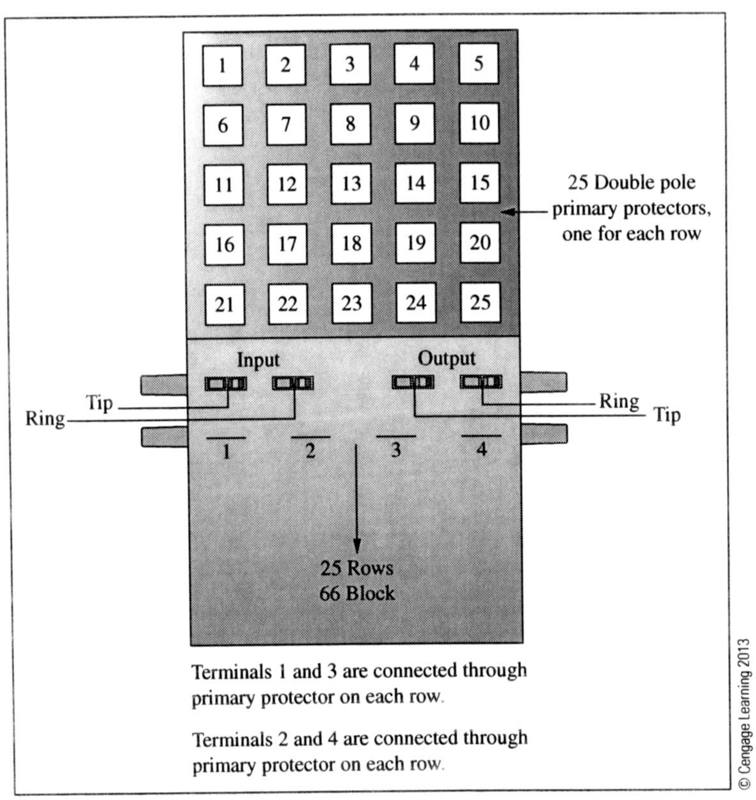

FIGURE 17–5

11. Use the DMM to test the block for continuity between lugs 1 and 3 and lugs 2 and 4.

12. Now, pull out fuse 1 and retest the continuity.
13. Replace the fuse.
14. Using the butt-set, locate the dial tone on the 66-block primary protector.
15. Confirm with the butt-set that you are connected to the correct line by calling the line simulator. The last digit in that telephone number indicates the simulator that you should be connected to. You should dial the number that you were issued and verify that the correct visual line indicator flashes.
    a. Example: If you were issued a number (555) 888-1232...you should be connected to simulator 2.
16. Install cross-connect wire from the output of the primary protector demark point to 66 block #2. Figure 17–6 illustrates the connection to 66 block #2. (Be sure that you are not punching down the cross-connect wires directly over the station jack wires. Only one wire should be attached to a 66-block lug at any one time. Always punch down on the lug directly next to the main connection lug, because they share continuity on the block.)

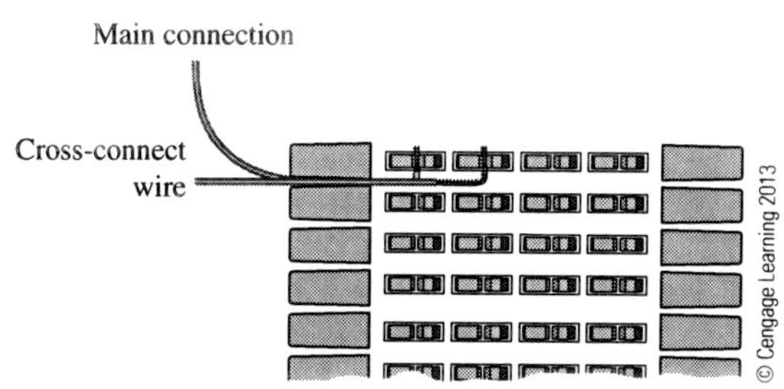

FIGURE 17–6

17. Rout the cross-connect wire around the cable routing spools.
18. At 66 block #2, cross-connect to the correct pair on the station jack. The station jack lines were installed during step 4 and step 7. Refer to them for correct wiring.
    a. In the United States, an 8-position (pin) jack is numbered 1 to 8, from left to right.
    b. Line 1 is always pins 4 and 5 in an 8-position, 8-conductor (8P8C) jack.
    c. On a 6P6C jack, the numbers run from 1 through 6, left to right.
    d. Line 1 is always pins 3 and 4 in a 6P6C (RJ-25) jack.
    e. On a 6P4C jack, the numbers run from 1 to 6, but there are no conductors in the number 1 or number 6 positions (RJ-14).
    f. On a 6P2C jack, the numbers again run from 1 through 6, left to right, but there are no conductors in pins 1 and 2 and pins 5 and 6 (RJ-11).
19. Connect a telephone handset to the station jack.
20. Pick up the telephone handset and test for dial tone.
21. Install the phone onto the station jack mounting bracket.
22. Once you've verified that you have dial tone on the station set, complete a second test by dialing back to the Ring It test phone. The loop should now be fully functional.

23. Ask the instructor to verify that the correct line was extended to the station jack and to review how each of the cross-connects were installed.
24. Clean up and complete the Final Questions.

FINAL SUMMARY

The previous lab procedure involved the installation of a traditional analog loop start phone system. The telephone company brings the line into the building and terminates it onto a primary protector to guard against lightning. From the primary protector, the line or lines are then terminated at the demarc, for the customer to make connection to their internal equipment. All installation past the demarc is the responsibility of the customer. The telephone company is only responsible for everything up to the demarc.

To complete an installation, a telecommunications technician is then required to be able to terminate telephone cables onto type 66 blocks, 110 blocks, patch panels, and wall outlets, as well as be able to make cross-connects between lines and station sets throughout the building. Understanding the pin-out of RJ-11, RJ-14, and RJ-25 jacks, along with correct wiring color codes, is critical to being able to install a complete system.

When lines become confused, the technician may also utilize a tone generator and tone tracer to track down unknown circuits. And finally, technicians use the butt-set to verify communication and functionality of a specific line before connecting station sets or making cross-connects. The butt-set, however, only provides basic line functions. Ultimately, the system must be tested with the actual station sets to verify all program features of the system, as developed by the manufacturer.

FINAL QUESTIONS

1. What is the significance of verifying dial tone and confirming that you are connecting to the correct circuit before extending the line?
2. Why are station cables not punched down directly over the top of connections from the demarc on the 66 block?
3. Explain any difficulty you had when performing punches.
4. Did you receive a dial tone on your first attempt? If not, explain what went wrong.
5. Why is it important to do a final test of the system with the actual customer station set instead of simply using the butt-set?

Name ____________________

Date ____________________

Class ____________________

LAB 18

Relays

REQUIRED READING

Introduction to Low Voltage Systems, 2nd edition, Chapter 12

INTRODUCTION

A *relay* is an electromagnetic switch, commonly used to control loads having large values of current and voltage. The low voltage industry uses relays for a variety of purposes, including fire alarm elevator capture, power supply supervision, and the programmable control of virtually any device. System control panels for fire alarm, security, and access-control typically provide a variety of relay outputs to be used for any number of purposes, depending on the complexity and design of the system.

Switch Contacts

The switch contacts for a relay are mounted on a spring-loaded metal plate. Control is provided by the internal coil, which, when energized, becomes an electromagnet. The energized electromagnet pulls in the metal plate, and the switch contacts change state. The relay is energized by applying the rated voltage to the coil. Relays can be made to operate from any level of AC or DC voltage and for any level of output power required by the system.

With no voltage applied to the coil, the switch contacts are in their normal state, and designated as either *normally open* (NO) or *normally closed* (NC). Relays are like mechanical switches in how they are configured. Each relay is chosen based on the number of poles and throws. The number of poles refers to the number of independent switches contained within the relay. The number of throws refers to the number of contact positions for each switch. Therefore, a single-pole double-throw (SPDT) relay has only one switch with two possible

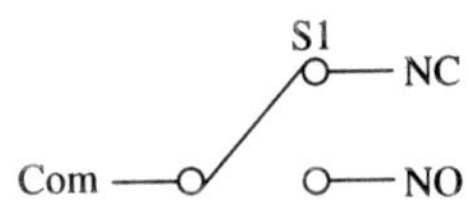

© Cengage Learning 2013

FIGURE 18–1

contact positions, one normally open and one normally closed. Figure 18–1 illustrates the normal state for a SPDT relay output.

Circuit Isolation

Relays also provide physical isolation between the low-voltage control side of the circuit and the high-voltage load. Figure 18–3 in the following procedure illustrates the schematic for a typical relay circuit. The dotted line indicates that the control between the coil and the switch contact is provided by an electromagnetic field. Except for the magnetic field, no physical connection exists between the control input and the output.

In the following lab procedure, the student connects a basic relay circuit to alternately control power to two LEDs, one red and one green. In addition, the student tests and measures the pull-in latching voltage and the release voltage for the relay coil.

PRE-JOB QUESTIONS

1. For the relay used in this experiment:
   a. What is the rated coil voltage?
   b. When energized, what is the rated coil current?
   c. What is the maximum voltage rating for the output contacts?
   d. What is the maximum current rating for the output contacts?
2. Explain the difference between normally open (NO) relay contacts and normally closed (NC) relay contacts.
3. What does *isolation* mean when talking about a relay circuit?

OBJECTIVES

- Connect and test a basic relay circuit.
- Test and measure the average pull-in latching voltage and the release voltage for a relay coil.
- Determine the configuration of the NO and NC switch contacts.

REQUIRED MATERIALS

1 double-pole double-throw (DPDT) 12-volt dc relay

2 LEDs: one red, one green

Two 330-ohm resistors

Hookup wire and wire strippers

PAD-234 analog digital trainer

TEST EQUIPMENT

Digital multimeter (DMM)

PROCEDURE

1. Obtain a double-pole double-throw (DPDT) relay with a 12-volt dc coil. Figure 18–2 illustrates the bottom pin-out of the relay.

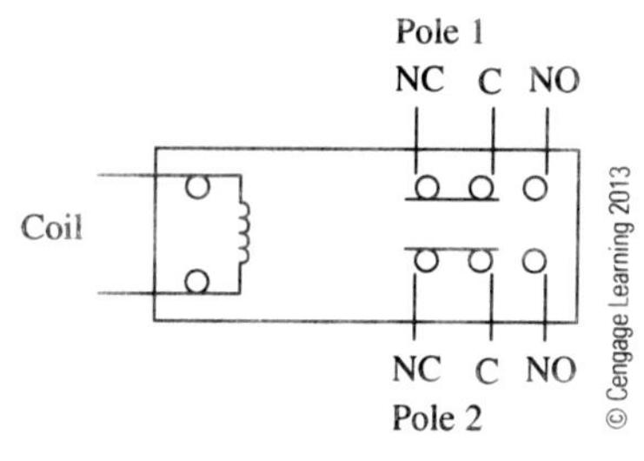

FIGURE 18–2 Double-Pole Double-Throw Relay.

2. Using the DMM, measure the coil resistance.

   Coil resistance = ____________ ohms (Ω)

3. The coil resistance is used to calculate the total current draw needed to pull in the relay. Calculate the coil current for a 12-volt power supply.

   I_{coil} ____________

4. Using the DMM, verify the NC and NO contacts. NC contacts should read zero ohms, and NO contacts should read infinite resistance.
5. Connect the circuit shown in Figure 18–3. Note: We are not really using a switch on the variable 12-volt power supply. To energize and de-energize the relay, simply connect and disconnect the positive side of the variable 12-volt supply when asked to do so.

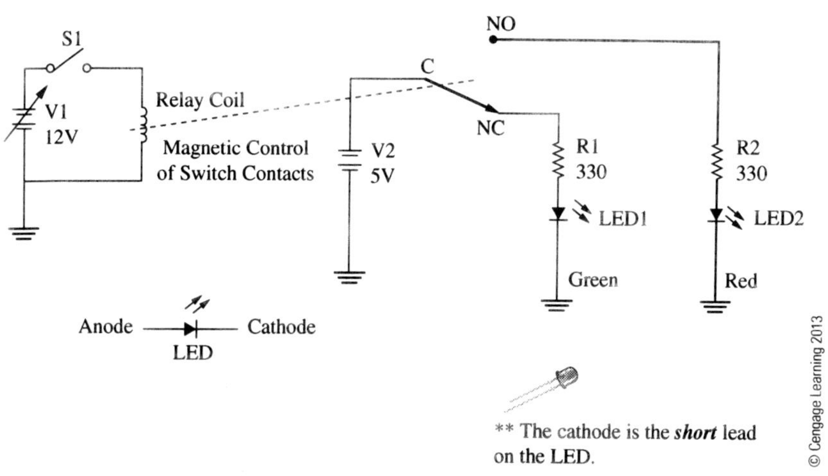

FIGURE 18–3

6. The *pull-in* voltage is the minimum value of coil voltage that causes the relay to turn on. To determine the pull-in voltage, turn the variable 12-volt control, V1, to its lowest setting. Connect the supply to the relay coil with a jumper wire. Slowly increase V1 until you hear the relay trip. It CLICKS when it powers ON. Measure and record the pull-in voltage in Table 18–1 as test 1.

TABLE 18–1

	Pull-In Voltage	Release Voltage
Test 1		
Test 2		
Test 3		
Average		

(c) Cengage Learning 2013

7. The *release voltage* represents the value of the coil voltage for when the contacts return to the unenergized position. Gradually lower V1 until the relay shuts off and the LEDs change state. Record the release voltage in Table 18–1 as test 1.
8. Repeat steps 6 and 7 two more times. Record the values in Table 18–1 as test 2 and test 3. Then average the values.
9. Which LED is ON when the relay is OFF? Which LED is ON when the relay is ON? Explain.
10. Connect the latching relay circuit, as shown in Figure 18–4. This time, use the fixed 12-volt supply.

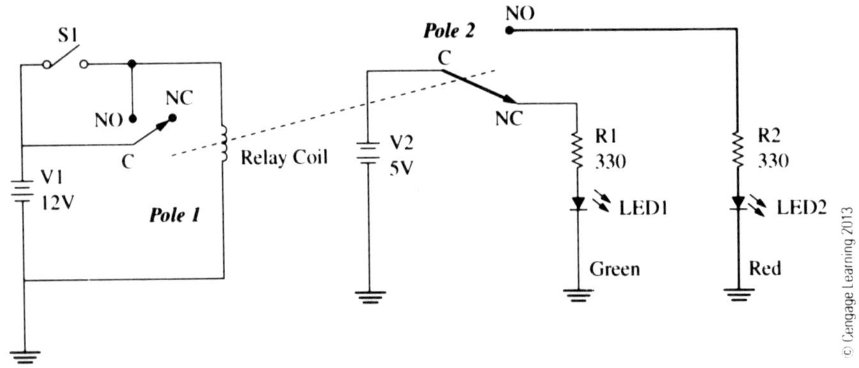

FIGURE 18–4 Latching Relay Circuit.

11. Be sure that switch 1 is disconnected. The green LED should be ON.
12. Connect switch 1 through a jumper wire. This applies power to the relay coil. Which LED is ON?
13. Disconnect switch 1 jumper wire. Which LED is ON? Explain.
14. How did the results of circuit 2 compare with those of circuit 1 when the relay was energized and de-energized?

FINAL SUMMARY

A *relay* is a magnetic switch with one or more sets of contacts used for controlling currents and voltages. The switches are controlled by an electromagnet called a *coil*. When the coil is de-energized, the *normally open* (NO) contacts are open with respect to the switch common (C), and the *normally closed* (NC) contacts are shorted to the switch common (C). Energizing the relay coil causes them to switch states.

The purpose of a latching relay circuit is to allow the relay to hold itself in the ON state after the initial energizing power to the coil has been removed. The latching circuit is commonly used in security and fire alarm circuits to continuously sound an alarm until the reset switch is opened.

FINAL QUESTIONS

1. Explain how an electromagnet is used in the design of a relay.
2. Do all relays operate at 12 volts dc, and do they all have the same coil current requirements? Explain.
3. Is the normally open contact of a relay open when the relay is energized or when it is powered down?
4. Come up with an idea for using a relay latching circuit. Draw the circuit below and explain how it works and its purpose.

Name ______________________

Date ______________________

Class ______________________

LAB 19 Security Systems

REQUIRED READING

Introduction to Low Voltage Systems, 2nd edition, Chapter 12

INTRODUCTION

It is recognized that an understanding of system devices, terminations, panel connections, and basic troubleshooting techniques are necessary skills for all limited energy installers. Understanding the theory of a security panel is greatly enhanced by having a working panel, along with input and output devices, for students to examine, connect, test, and troubleshoot. The following lab procedure outlines the connections, function, and testing of a basic security system.

It is not the intent of this exercise to teach a specific type of security system. Although the type of system used in this exercise is the DSC PC1832, with minor modifications to the lab procedure any type of basic security panel can be used for this training exercise.

System programming is not covered in this lab procedure. Therefore, it is the responsibility of the instructor to perform the initial system setup and programming before attempting the exercise. The instructor must also address any unique manufacturer specifics to ensure that the system functions as required. For instructor reference, programming of the DSC PC1832 can be accomplished directly through the keypad or through a PC with a network communication link or through an RS232 interface. To accomplish the PC communication and programming options, additional software and drivers must be downloaded from the manufacturer.

Before doing this lab exercise, it is recommended that the security panel be mounted on a piece of ¾-inch plywood and that all internal panel connections be wired to external terminal strips, as shown in Figure 19–1, and labeled. To do so helps prevent any undo wear and tear on the internal connections of the control panel.

In addition, system devices are also mounted on the plywood, to each side of the security panel, and in like manner wired to external terminal strips. To complete this lab procedure, students wire between the external device terminal strips and the external control panel terminal strips in order to make the necessary connections.

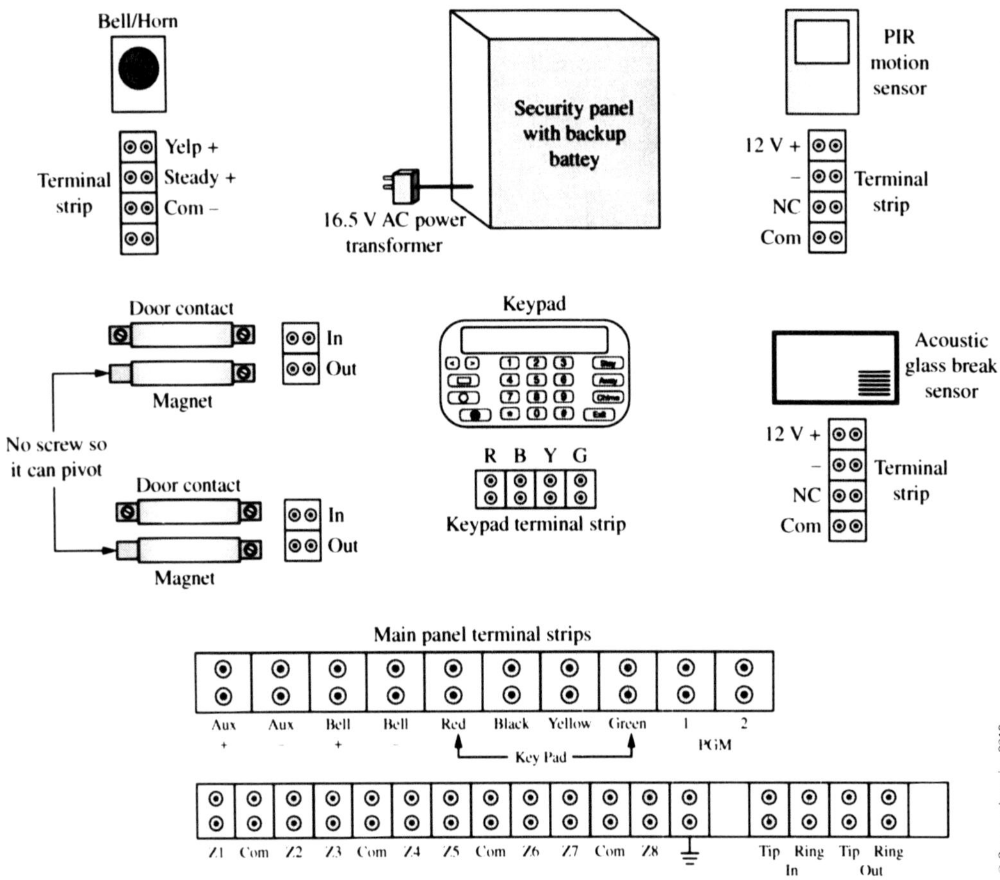

FIGURE 19–1

A main ac power switch and a dc battery power switch are also needed in order to connect and disconnect system power as required.

Zones for the upcoming lab procedure are to be programmed by the instructor as follows:

Zone 1	Front door	NO door contact	Single end-of-line connection
Zone 2	Back door	NO door contact	Single end-of-line connection
Zone 3	PIR (passive infrared) motion detector	NC	Single end-of-line connection
Zone 4	Acoustic glass break sensor	NC	Single end-of-line connection
Zones 5–8	Not used		

All zones require closed loops to operate. An open loop on any zone is detected as a trouble alert. When the panel is armed, a zone transitioning from closed to open triggers an alarm.

PRE-JOB QUESTIONS

1. What is the difference between closed loop and open loop?
2. What is the value of the end-of-line (EOL) resistor you are using for this lab?
3. What is the purpose of a tamper switch?
4. In the space provided here, draw a single EOL connection having one NO and one NC device attached to a single zone.

5. In the space provided here, draw a double EOL connection having one NC device attached to a single zone.

OBJECTIVES

- Install the basic components of a security system, including motion sensor, acoustic glass break sensor, and door contacts.
- Recognize and demonstrate the installation differences between a single EOL and a double EOL connection.
- Test the system for functionality.
- Calculate the required amp-hours (Ahr) for a standby battery, based on total current draw.

REQUIRED MATERIALS

DSC PC1832 security panel
Specification sheets for the security panel
16.5-volt, 40 VA ac power supply
PK 5500 keypad
PIR motion sensor
Acoustic glass break sensor
Acoustic glass break sensor tester
2 sets of door contacts, NO closed loop with magnet connected
Bell with steady or yelp option
12-volt, 7-Ahr backup battery
2 pairs 24 AWG wire
Spade connectors and crimp tool
Eight 5.6-kiloohm (kΩ) resistors

TEST EQUIPMENT

Digital multimeter (DMM)

PROCEDURE

1. Be sure that the ac power switch to the panel is OFF and that the dc battery switch on the control panel is OFF.
2. Locate the PK5500 keypad terminal block and the main terminal block for the control panel.
3. Connect the four keypad terminal block wires (red, black, yellow, green) to the control panel terminal block, using the crimp-style spade connectors.
4. Power up the panel by turning the main ac power switch ON.
5. Look for troubles on the keypad indicator (battery off, bell trouble, etc.). Use the keypad scroll arrows to scroll through the trouble indicators.
6. Turn on the battery power switch. The trouble indicator should clear.
7. Wire the bell to the bell output of the main terminal block. Be sure to observe proper polarity when making the connection. Connect the +bell terminal to the Yelp or Steady, as desired.
8. Use the scroll arrows on the keypad to check the trouble function indicators.
9. Reverse the leads on the bell, swapping the + and the − terminals.
10. What happens if it is wired backward?
11. Rewire the bell in the forward direction.
12. Check the keypads to see which zones are open. Use the keypad scroll arrows to scroll through the open zones.

13. Using a DMM, measure the voltages of the open zones at the main terminal block.

    Open zones = ____________ volts

14. Terminate zones 5, 6, 7, and 8 by placing a 5.6-kΩ resistor across each of the input terminals and common. This dummies out the unused zones and prevents the control panel from thinking that they are open.
15. Remeasure the voltages on zones 5, 6, 7, and 8.

    Terminated zones = ____________ volts

16. Locate the door contacts. The door contacts are normally open (NO); this means that they act as an open switch when the magnet is not pulling in the contacts.
17. Using an ohmmeter, measure the resistance of the door contacts with the magnet turned 90° from the contact. Then measure the resistance of the contact when the magnet is parallel with the door contact. (Remember: Magnetic fields that are turned 90° will not be detected.)

    Turned 90° ____________ Ω Parallel ____________ Ω

18. Determine from Figure 19–2 how the single EOL resistors must be connected when the door contacts are properly installed. Should resistors be connected in series or in parallel with the NO door contacts? __________

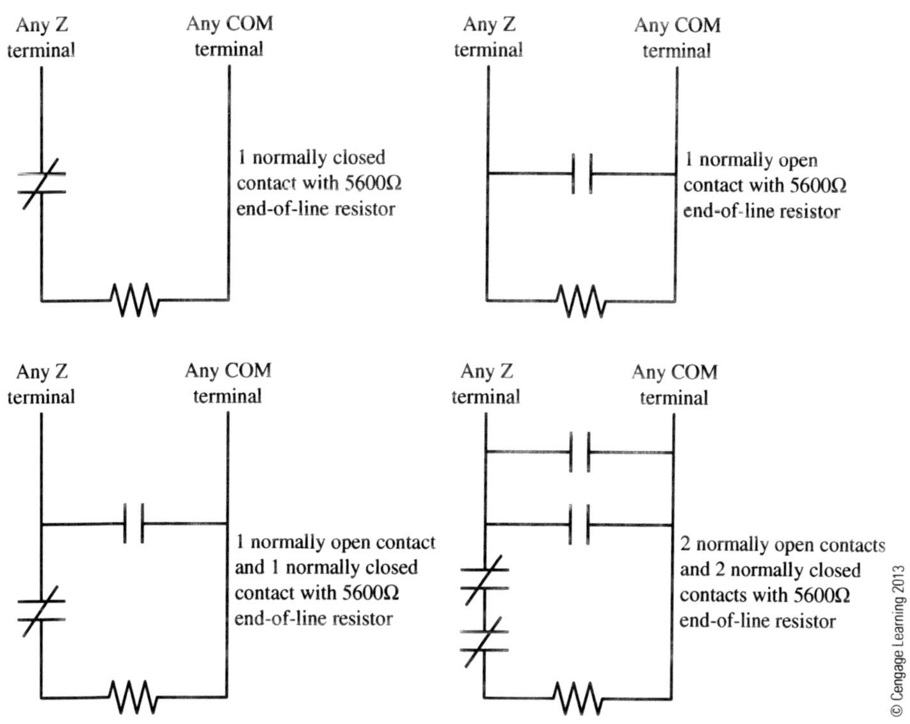

FIGURE 19–2

19. Verify with your instructor the connection you have chosen for the EOL resistors before making the connection.
20. Connect the door contacts and EOL resistors to zones 1 and 2. Zone 1 represents the front door, and zone 2, the back door.
21. Measure the voltage of zone 1 and zone 2 to common with the door contacts in place and engaged.

Zone 1____________ volts Zone 2 ____________ volts

22. Verify the function of the door contacts by turning the magnet 90° degrees and measuring the voltage at zone 1 and zone 2.
23. Replace the magnets to their engaged position.
24. Locate the PIR motion sensor and the PIR motion sensor terminal strip.
25. Wire the 12-volt dc power from the Aux+ and Aux− output terminals on the main terminal strip to the power input terminals on the PIR motion sensor terminal strip.
26. Because the PIR motion sensor uses an NC contact, determine how the EOL resistor is to be connected for a single EOL connection.
27. Connect the EOL resistor and NC contacts of the PIR motion sensor to zone 3 of the main terminal strip.
28. Measure the voltage of zone 3 to common.

Zones 3____________ volts

29. Check the function of the PIR motion sensor by covering the detector window with a piece of black tape. When covered, the device should change state, and the zone voltage should change to that of an open zone.
30. Locate the acoustic glass break sensor and the acoustic glass break sensor terminal strip.
31. Wire the 12-volt dc power from the Aux+ and Aux− output terminals on the main terminal strip to the power input terminals on the acoustic glass break sensor terminal strip.
32. Because the glass break sensor uses an NC contact, determine how the EOL resistor is to be connected for a single EOL connection.
33. Connect the EOL resistors and acoustic glass break sensor to zone 4 of the main terminal strip for a single EOL connection.
34. Measure the voltage of zone 4 to common.

Zones 4____________ volts

35. Have your instructor check the function of the acoustic glass break sensor by using the acoustic glass break tester. If the device is set to momentary mode, it should reset by itself within 30 seconds. If the device is in latching mode, then you need to cycle the power to get it to reset. When activated, the circuit should open, and the voltage measurement should indicate an open zone.

36. Check the keypad for any trouble alerts (everything should now be set).
37. Have your instructor give you the disarm pass code. ____________
38. Arm the panel by pressing the AWAY button for 2 seconds.
39. Wait until the panel is armed.
40. Move one of the door contact magnets 90º. The panel should go into alarm, and the bell output should sound.
41. Enter the pass code to disarm the system
42. Power down the system, disconnect all wires from the terminal strips, and clean up the lab area.

FINAL SUMMARY

Classification of Power Supply and Cabling

The power supply and cabling for a low-voltage security system is rated as Class 2 and installed according to *Article 725* of the *NEC*.

NC versus NO and Zone Connections

Most security sensors operate from an NC state and open during an alarm condition. Some devices, however are configured NO and close during an alarm condition. An example of an NO device would be a smoke detector. NO devices are installed in parallel with the EOL resistor. When triggered, the NO device changes state and shorts the EOL resistor, thereby forcing the panel into alarm.

Maximum Current Rating of the Panel

Aside from being able to connect input sensors and output devices to a control panel with the appropriate EOL resistor connection, an installer is also required to calculate the maximum current draw of a system to guarantee that the control panel and power supply are capable of providing the necessary power. Every security panel will have a maximum limit as to how much current the main panel board can provide to connecting devices, sensors, and peripheral add-on modules. For the DSC PC1832, the panel can provide up to a maximum of 700 mA of current flow. If more current is required, an additional power supply is needed. Always be sure to check the electrical specifications for every device, sensor, module, and keypad connected to the panel. Add up the total current draw, and make sure it does not exceed the maximum limit for the system.

Auxiliary Outputs

Auxiliary outputs are also limited as to the total amount of current they can individually provide to a device. Always check specification data sheets for output current maximums, to guarantee that terminals are not overloaded when connecting devices; this is especially critical when connecting relays and devices from other manufacturers.

Standby Power Capacity

A sealed, rechargeable, lead acid or gel-type battery is required to meet UL requirements for standby power. UL residential and commercial burglary systems require a minimum of 4 hours of standby power. UL/ULC residential fire and healthcare installations require a minimum of 24 hours of standby power. And ULC (Underwriters Laboratories) commercial fire and burglary installations require a minimum of 24 hours of standby power.

As an example, a 4-Ahr battery can supply 4 amps of current for up to 1 hour. Reducing the current draw to 700 mA means that the same battery can supply current for up to 5.7 hours. Be aware that batteries will deteriorate over their natural life span, so the previous calculation is only true for new batteries. To guarantee power for the required time, it is safer to use a 70 percent adjustment factor: 5.7 hours x 0.7 = 4 hours. With the 70 percent factor calculated in, it is shown that the 4-Ahr battery can reliably provide 700 mA of current for up to 4 hours. Battery capacity will deteriorate over time, and it is recommended that they be replaced every 3–5 years.

FINAL QUESTIONS

1. A low-voltage security system is installed using Class 2 cable and Class 2 wiring methods. Which article of the *NEC* is used to determine the correct installation requirements for Class 2?
2. Explain the difference between an NO contact and an NC contact. How can you determine which type you have? And how are they connected?
3. When installing a relay to the auxiliary output of the security panel, or devices that have not been specified by the manufacturer, what considerations must be taken into account before choosing the type of device?
4. Calculate the required Ahr rating for a standby battery if the total current draw for the system is 500 mA and you want to supply 24 hours of standby power. Use a 70 percent safety factor when making the calculation.

Name ______________________________

Date ______________________________

Class ______________________________

LAB

20

Access Control Systems

REQUIRED READING

Introduction to Low Voltage Systems, 2nd edition, Chapter 12

INTRODUCTION

It is recognized that an understanding of system devices, terminations, panel connections, and basic troubleshooting techniques are necessary skills for all limited energy installers. Understanding the theory of an access control panel is greatly enhanced by having a working panel, along with input and output devices, for students to examine, connect, test, and troubleshoot. The following lab procedure outlines the connections, function, and testing of a basic access control system.

It is not the intent of this exercise to teach a specific type of access control system. Although the type of system used in this exercise is the KeriSystems PXL-500, with minor modifications to the lab procedure any type of access control panel can be used for this training exercise.

System programming is not covered in this lab procedure. Therefore, it is the responsibility of the instructor to perform the initial system setup and programming before attempting the exercise. The instructor must also address any unique manufacturer specifics to ensure that the system functions as required. For instructor reference, programming of the PXL500 can be accomplished by a PC through an RS232 interface or LAN connection. To accomplish the PC communication and programming options, additional software and drivers must be downloaded from the manufacturer.

Before doing this lab exercise, it is recommended that the access control panel and power supply be mounted on a piece of ¾-inch plywood and that all internal panel connections be wired to external terminal strips, as shown in Figure 20–1, and labeled. To do so helps prevent any undo wear and tear on the internal connections of the control panel.

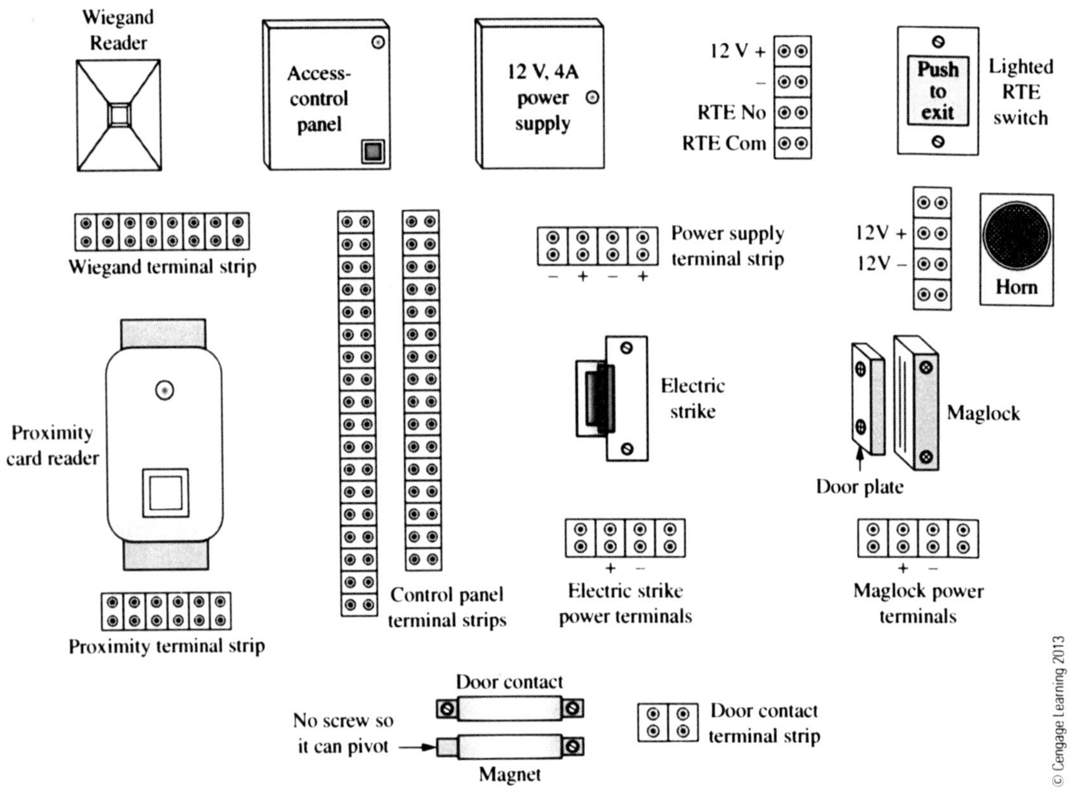

FIGURE 20–1

In addition, system devices are also mounted on the plywood, to each side of the access control system, and in like manner wired to external terminal strips. To complete this lab procedure, students wire between the external device terminal strips and the external control panel terminal strips in order to make the necessary connections.

For the maglock, it is best to build a miniature door into the lab display. Another option would be to mount the door plate on a spring-loaded assembly, as shown in Figure 20–2.

FIGURE 20–2

A main ac power switch is needed in order to connect and disconnect system power as required.

PRE-JOB QUESTIONS

1. Electronic access control requires human intervention. True or false?
2. What are the three most common types of cards used in an access control system?
3. Explain how a Wiegand card differs from a magstripe card.
4. Besides access cards, what other types of credentialing can be used with an access control system?
5. What is meant by the acronym RTE?
6. Describe the difference between a fail-safe lock and a fail-secure lock. What type of lock will you be using in this lab?
7. What is the purpose of a transient protector when installing electric strikes and maglocks?
8. What type of sensor is used to detect that a door has been left open?
9. Give two examples of an output device for an access control system.

OBJECTIVES

- Connect and test an access control panel.
- Demonstrate the connection of power and transient protection for a fail-safe or fail-secure lock.
- Demonstrate the connection of card readers to the access control panel.
- Demonstrate the connection of RTE switches and door contacts.
- Demonstrate the connection of an alarm siren to the access control panel.

REQUIRED MATERIALS

KeriSystems PXL-500 access control panel
12-volt dc, 4-amp power supply for main panel, locks, and siren
Safety glasses
22/4 AWG wire
Wire strippers
Proximity card reader
Wiegand card reader
Proximity cards
Wiegand cards
Electric strike (12-volt dc, fail-secure)

Maglock (12-volt dc)

Transient protectors (3)

Lighted RTE switch (normally open, momentary switch, 12-volt dc)

Magnetic contacts (NO) (closed-loop when magnet is engaged)

Alarm siren/horn (12-volt dc)

TEST EQUIPMENT

Digital multimeter (DMM)

PROCEDURE

Part I

1. Do not power up the panel without instructor permission.
2. Wire the electric strike to lock 1 through the 12-volt power supply, as shown in Figure 20–3. The electric strike controls door 1. Be sure to connect the transient protector in parallel with the lock.

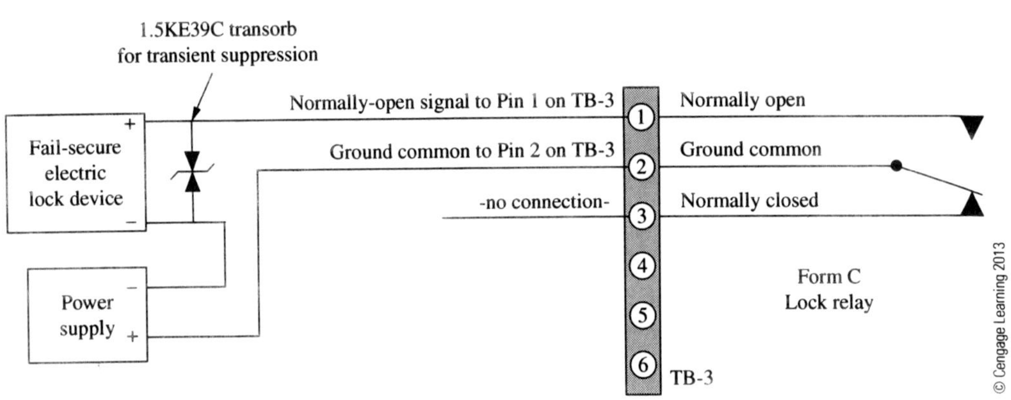

FIGURE 20–3

3. The maglock controls door 2 (all maglocks are fail-safe). Wire the maglock to lock 2 through the 12-volt power supply, as shown in Figure 20–4. Be sure to connect the transient protector in parallel with the lock.

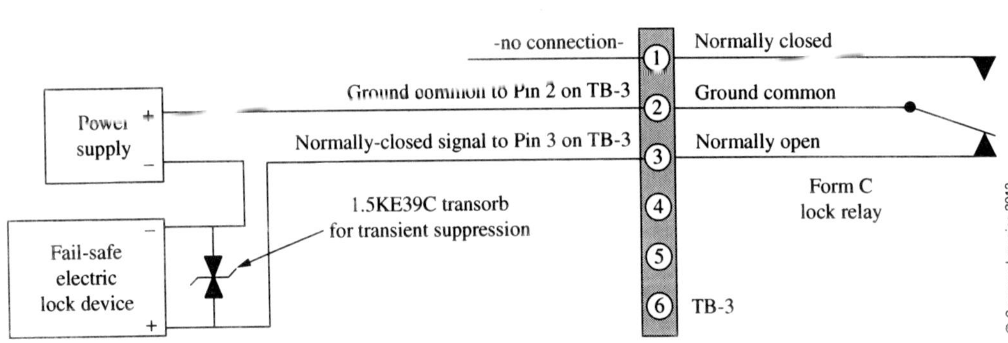

FIGURE 20–4

4. Locate proximity reader A for door 1.
5. Wire proximity reader A to the control panel for door 1, as shown in Figure 20–5. Connections include reader antenna, beeper, power, ground, red LED, and green LED.

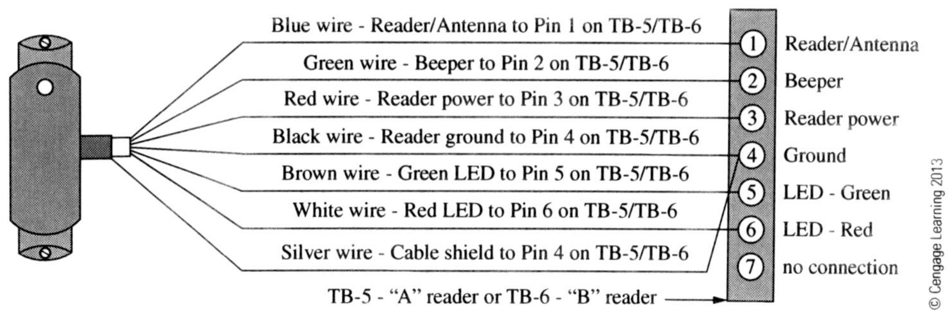

FIGURE 20–5

6. Locate the Wiegand reader B for door 2.
7. Wire Wiegand reader B to the control panel for door 2, as shown in Figure 20–6. Connections should include data 0, data 1, beeper, power, ground, red LED, and green LED.

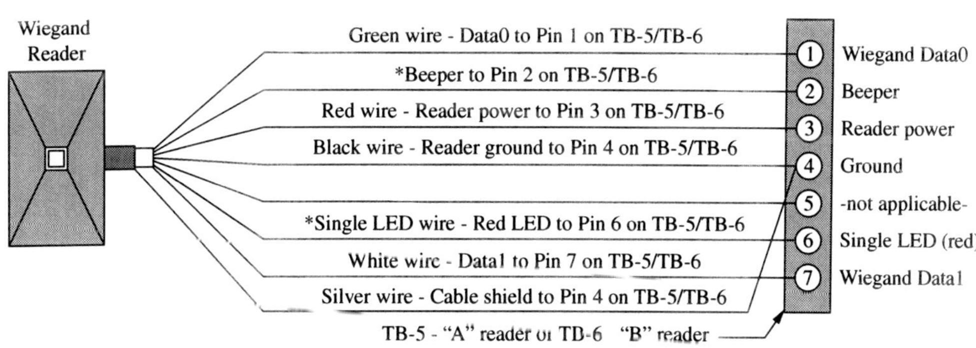

*Refer to the Wiegand Reader's documentation for the color of this wire.

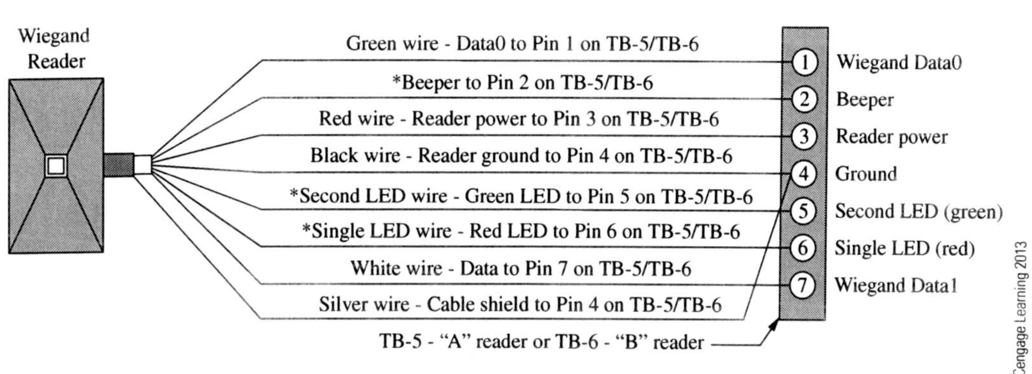

*Refer to the Wiegand Reader's documentation for the color of this wire.

FIGURE 20–6

8. Insert a shorting jumper to the door switch input for door 1. Later in the exercise, a magnetic door contact is connected to this input.
9. Have your instructor check off your wiring. ___________
10. Switch the power ON.
11. Test proximity reader A by reading a card. Verify that the electric strike releases.
12. Test Wiegand reader B by reading a card. Verify that the maglock releases.
13. Switch the power OFF.

Part II

14. Using the digital multimeter, verify that the request-to-exit (RTE) switch is normally open (NO), and momentary.
15. Wire the RTE switch to the RTE input for door 2.
16. Have your instructor verify your wiring. ___________
17. Switch the power ON.
18. Engage the maglock and test the RTE switch. Make sure the maglock releases.
19. Switch the power OFF.

Part III

20. Using the digital multimeter, verify that the magnetic door contact is a NO variety. The switch will close when the magnet is placed in proximity to the switch.
21. Remove the shorting jumper, and connect the magnetic contact to the door switch input for door 1.
22. Wire the alarm siren and 12-volt power supply to the NO relay output for door 1, as shown in Figure 20–7. Be sure to connect the transient protector in parallel with the siren.

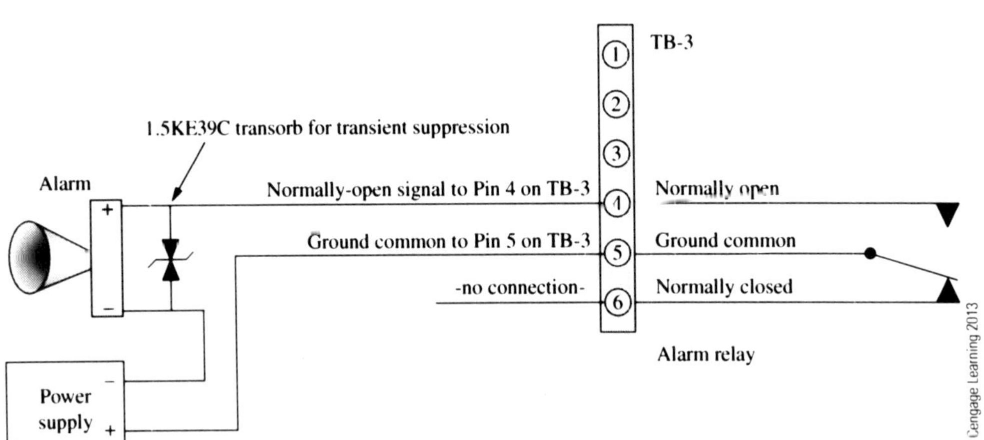

FIGURE 20–7

23. Have your instructor verify your wiring. __________
24. Switch the power ON.
25. Unlock door 1 by reading a card on proximity reader A.
26. Open the magnetic door contact by moving the magnet away from the switch.
27. Leave the magnet open for the determined time after the lock reengages to test the alarm siren.
28. Verify that the alarm siren energizes. The siren activates to notify security that the door has been propped open.
29. Once verified, close the magnetic contact, and reset the panel alarm.
30. Power down the panel.
31. Remove all wiring and clean up the lab area.

FINAL SUMMARY

Before the availability of electronic access control, access control involved a guard station and human intervention to verify an individual's credentials. Once verified, a door release switch was pressed at the guard station to momentarily unlock the door when granting access to a secured space.

For modern electronic access control systems, human intervention is not required, because the control panel makes uses of electronic credentialing for granting access. Electronic access control inputs involve types of credentialing systems, such as card readers, keypads, and biometrics, as well as magnetic door contacts and request-to-exit (RTE) switches. Depending on the type of card reader, keypad, or biometric used, wiring instructions will differ. Always be sure to follow the manufacturer's instructions.

Locks require a separate power supply to be sure there will be enough available power to control the lock without potentially loading down the panel. As demonstrated in this lab, the installation differences between a fail-secure and a fail-safe lock will be in how they are wired to the output relays of the control panel.

Transient protectors, or transorbs, are used to protect the control panel from any potential damaging inductive kick when the locks or sirens are shutting down. Without transient protection, a collapsing magnetic field could generate momentary surges in the range of thousands of volts, which could potentially destroy the control panel and all connecting devices.

Input devices include card readers and biometrics.

Output devices include lights, maglocks, electric strikes, sounders, sirens, alarms, horns, strobes, and buzzers.

FINAL QUESTIONS

1. How did wiring the Wiegand reader compare to wiring the proximity reader?
2. How did wiring the fail-secure lock compare with the wiring of the fail-safe?
3. Why is a separate power supply needed when installing locks?
4. Explain how a video camera added to an access control system could be used.
5. What should be your first step when installing an access control system or any unknown devices?

Name ______________________

Date ______________________

Class ______________________

LAB 21 Installation of HD TV Antenna

REQUIRED READING

Introduction to Low Voltage Systems, 2nd Edition, Chapter 13

INTRODUCTION

Choosing the Right Antenna

Locally broadcast HDTV (high definition television) in the United States is delivered over the UHF band of frequencies. To find the best type of HDTV antenna for your area, go to the following web address:

http://www.antennaweb.org

The Consumer Electronics Association (CEA) has created this Web site to help people determine the best type of outdoor HDTV antenna to purchase for their area of the country. Once you're at the Web site, enter your zip code and click Submit. The information returned provides you with the best antenna options available for your location, based on the number of broadcast frequencies and channels in your area. The Web site also provides a map showing where you live in relationship to the closest transmitting towers, information you will need while tuning and peaking the signal during the installation process.

If you purchase a CEA-certified outdoor antenna, it will come with a color code. Be sure the color code matches the color code provided to you from the Web site. If you need to receive signals from multiple transmitter stations that are rated for different colors, then you need to purchase a more expensive antenna, one that can receive a wider band of frequencies and is rated for multiple colors.

Installing a basic outdoor antenna involves the following steps:

- Installing the mast support and mast
- Mounting the antenna to the mast

- Connecting the coaxial cable to the antenna feed block
- Grounding the antenna mast and coaxial cable
- Pulling the coaxial cable into the building or residence
- Connecting the antenna feed to the receiver
- Tuning and peaking the reception

Antenna splitters and signal amplifiers will be covered in the final summary.

PRE-JOB QUESTIONS

1. What is a *balun*?
2. What article of the *NEC* covers the installation of radio antennas?
3. What minimum size copper bonding or grounding electrode conductor is required to ground an antenna?
4. Aluminum grounding electrode conductors shall not be installed within __________ inches of the earth.
5. Antenna lead-in conductors shall maintain a minimum clearance of __________ feet from electric light and power service conductors of 250 volts or less.
6. Can an antenna or antenna lead-in conductors be attached to an electrical service mast?

OBJECTIVES

- Determine the best type of antenna for your area of the country.
- Demonstrate the installation of the antenna mast.
- Mount the antenna to the mast and determine optimal signal direction.
- Demonstrate antenna system grounding.
- Demonstrate how to peak the antenna for maximum gain.
- Determine the amount of loss or gain for signal splitter or amplifier.

REQUIRED MATERIALS

HDTV digital outdoor antenna

Antenna mast, 1½-inch (38.1-mm) pipe

Mast support base for roof mount

Mounting hardware

HDTV with built-in ATSC tuner for HDTV reception

75-ohm outdoor all-weather coaxial cable (enough to run from the antenna location to the grounding block)

A shorter length of 75-ohm, outdoor all-weather coaxial cable to run from grounding block to the television receiver

Coaxial cable grounding block
10 AWG copper grounding conductor
3 ground clamps (one for the antenna mast and two for the ground rod)
Silicone

TEST EQUIPMENT

Antennas Direct SM200 HDTV signal finder

If you do not have an HDTV signal strength meter, you can use the meter included in the setup menu of the HDTV receiver.

PROCEDURE

Determine What Type of Antenna to Use

1. Go to http://www.antennaweb.org to determine what type of antenna to purchase for your area of the country.
2. Enter your zip code, and click Submit.
3. What color code is given for the antenna to be used in your area?

   ____

4. Print the map so you can determine the location of the transmitting towers in your area.

Install the Antenna

5. Choose a location for the antenna that is free from any obstructions or overhead wires. Be sure you have a clear line of sight in the direction of the transmitting tower.
6. Install the support base for the antenna mast on the roof of the building, using appropriate hardware.
7. Attach the antenna mast to the support base, and bolt it into place.
8. Attach the ground clamp to the antenna mast near the base.
9. Prepare the antenna; this may require unfolding antenna elements. Be careful not to bend them in the process.
10. Locate the antenna mast clamp and U-bolts.
11. Place the antenna on the top of the antenna mast, and point the front of the antenna in the general direction of the transmitter tower.
12. Loosely tighten the U-bolts to hold the antenna in place; they will be tightened after the signal is peaked for maximum gain.

Attaching the Cables

13. Attach an F-connector to one end of an outdoor-rated, 75-ohm coaxial cable. Refer to Lab 2 of this manual for attaching the connector.

14. Attach the coaxial cable to the antenna lead-in block.
15. Support the coaxial cable along the length of the antenna mast with cable ties, being sure to leave enough slack at the top. Be careful not to pull the cable ties too tightly, or you could potentially damage the cable.
16. Attach a 10 AWG grounding conductor to the ground clamp near the base of the antenna mast.
17. Run the 10 AWG grounding conductor and the coaxial cable down the side of the building to the location designated as the service entrance.

    (An 8-foot (2.4-m) ground rod should be installed just under the service entrance unless the electrical service ground already happens to share the same location. If a separate communication ground rod is used, it must be bonded to the electrical service with 6 AWG copper.)

Installation at the Point of Entrance

18. Do not cross any electrical service conductors with the coaxial cable, and be sure to maintain a minimum distance of 2 feet (0.6 m) from all electrical service conductors measuring less than 250 volts to ground.
19. Locate the coaxial cable ground block.
20. Attach the coaxial cable feed from the antenna to the coaxial cable ground block.
21. Locate the coaxial cable that is to be pulled into the building and attach an F-connector to the end.
22. Attach the coaxial cable to the other side of the ground block.
23. Mount the ground block to the side of the building, about 1 foot (0.3 m) below the entry point for the coaxial cable.
24. Run the coaxial cable inside the building to the back of the HDTV receiver, and seal up the entry point of the building with silicone. Be sure to leave a drip loop outside the building for future maintenance.

Grounding the Antenna

25. Attach two ground clamps to the communication ground rod or to the electrical service ground rod.
26. Attach the 10 AWG grounding electrode conductor from the antenna mast to one of the ground clamps, and tighten.
27. Cut a piece of 10 AWG copper to attach between the communication ground rod and the coaxial cable ground block.
28. Attach one end of the 10 AWG to the other ground clamp, and the other end to the coaxial cable ground block.
29. Tighten the ground block screw.
30. The completed assembly should look like Figure 21–1.

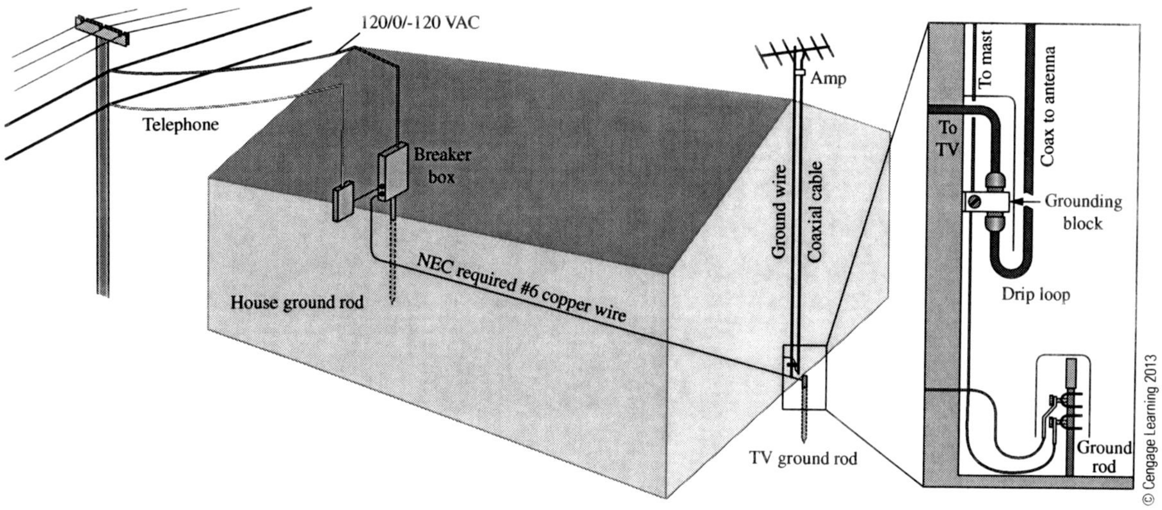

FIGURE 21–1

Setting up the Receiver

31. Attach an F-connector to the end of the coaxial cable at the back of the television receiver, and connect it to the antenna input.
32. Turn on the HDTV receiver and press Menu.
33. Select antenna input.
34. Locate channel preset, and press auto load to load all available channels into the receiver.
35. Select a channel for viewing.

Peaking the Signal at the Antenna

36. Disconnect the coaxial cable feed to the antenna, and attach a short piece of coaxial cable between the antenna block and the antenna input on the SM200 HDTV signal finder.
37. Attach the disconnected cable feed to the receiver side of the meter.
38. Loosen the antenna mount, and rotate the antenna on the horizontal plane until maximum signal gain is indicated on the meter.
39. Tighten down the U-bolts on the antenna mast to lock in the position of the antenna.
40. Disconnect the meter and remove the short piece of coaxial cable from the antenna block.
41. Reattach the main coaxial cable feed to the antenna block.

Peaking the Signal from the Receiver

42. Go into the setup menu of the HDTV receiver, and locate the signal strength meter.
43. The next step will take two people to complete: one person at the receiver and the other person at the antenna.

44. While communicating over cell phone, have the person at the antenna adjust the horizontal rotation of the antenna until the signal meter peaks out to a maximum.
45. Check additional channels on the receiver to be sure that the signal is peaked.
46. Tighten down the U-bolts on the antenna mast to lock in the position of the antenna.
47. The antenna is now installed.

FINAL SUMMARY

UHF channels occupy the frequency band between 300 MHz and 3 GHz. Channels 14–69 fall between 470 MHz and 806 MHz. A normal signal level at a receiver should measure between 10 and 15 dB.

Splitters and Amplifiers

The installation of multiple antenna feeds throughout a house or building requires the use of splitters, baluns, and signal amplifiers. A 2:1 split reduces the signal level at the receiver by half, which theoretically will produce a 3-dB loss for a perfect splitter. The reality, however, is that most splitters and baluns are only 80 to 90 percent efficient. The expected loss will therefore be between 3.5 and 4 dB. The list below indicates the expected losses for a variety of splitters and a 75-ohm to 300-ohm balun.

2:1 splitter	3.5–4 dB loss
3:1 splitter	5–6 dB loss
4:1 splitter	7–8 dB loss
75 ohm to 300 ohm balun	0.2–2 dB

At some point, splitting will affect the picture quality by reducing the signal-to-noise ratio. Amplification can help, but the question becomes where to introduce amplification. Amplifying a poor signal amplifies the noise as well, and there will not be any noticeable gain. However, amplifying a signal too early could possibly result in over-amplification and distortion. The following list indicates levels of amplification and signal loss and how they are relate to dB.

20 dB = a gain of 100
10 dB = a gain of 10
3 dB = a gain of 2
0 dB = no gain
−1 dB = a 20% loss of signal
−3 dB = a 50% loss of signal
−10 dB = a 90% loss of signal

Receivers have built-in amplifiers as well as auto gain control (AGC). The AGC sets the signal level for all channels to an equal value. To prevent over-amplification, signal amplifiers should only be used to compensate for cable losses and losses due to over-splitting.

A general rule of thumb is that RG-6 will lose 1 dB of signal every 18 feet (5.5 m) for channel 52. If digital dropout is noticed at the receiver, then a signal amplifier mounted at the antenna mast is typically required. However, do not overamplify, or distortion may occur.

Mast-mounted amplifiers are connected in-line with coaxial cable at the antenna feed. The dc power to the amplifier is applied through the central conductor of the coaxial cable. Voltages may vary between 12 V and 20 Vdc, depending on the manufacturer.

When feeding multiple rooms to a building or residence, a distribution amplifier is preferred, especially when cable feeds exceed 150 feet (45.7 m). Distribution amplifiers provide multiple outputs, thereby reducing the need to use splitters.

Attenuators

If signal levels are too high at the receiver, an attenuator may be used to help reduce the level. Attenuators are available that reduce signal levels by as much as 3 to 10 dB. Attenuators are passive filter networks that screw onto the end of the coaxial cable before attaching to the receiver.

FINAL QUESTIONS

1. What would be the expected signal loss for 100 feet (30.4 m) of RG-6 cable connecting through a 2:1 splitter? Use channel 52 as reference.
2. If a separate communications ground rod is installed, does it need to be bonded back to the main electrical ground rod?
3. When is the use of a distribution amplifier preferred over a splitter?
4. How is power applied to an antenna-mounted signal amplifier?

Name ______________________

Date ______________________

Class ______________________

LAB 22

Camera Systems

REQUIRED READING

Introduction to Low Voltage Systems, 2nd edition, Chapter 14

INTRODUCTION

When designing a CCTV (closed circuit television) camera system, the first order of business is to define the space. In doing so, the following items need to be determined:

- Requirements of the space
- Size of space measured in length × width × height
- Distance from the camera to the subject
- Type of lighting within the space over a 24-hour period of time
- Number of pixels per foot required

The type of camera to be used can then be determined based on the analysis of the above data.

In this lab exercise, students design a CCTV system for a given space and determine the minimum requirements for camera specifications. The lab procedure will walk them through the process.

PRE-JOB QUESTIONS

1. What is the width in millimeters of a ¼-inch camera sensor? __________ mm
2. 1 lux = __________ foot-candles.
3. How many pixels per foot are required for facial recognition?
4. What is the formula to calculate horizontal field of view?
5. How do you convert horizontal field of view to vertical field of view?

OBJECTIVES

- Calculate horizontal and vertical fields of view for a given space.
- Determine the type of camera to be used for a given space.
- Determine lens specifications.
- Determine the lighting requirements over a 24-hour period of time.
- Determine the digital camera resolution based on minimum required pixels-per-foot.

REQUIRED MATERIALS

Calculator
Tape measure

TEST EQUIPMENT

Light meter

PROCEDURE

The purpose of this lab exercise is to help you determine what type of camera and lens to select for a given space. You can choose any type of space; the only requirement is that you are limited to one camera with a fixed iris lens, and it cannot be an infrared camera. It must operate within the existing light of the space.

1. Draw your space in the box below, and determine where you plan to mount the camera.

2. Using a tape measure, measure the length, width, and height of the space.

   Length ________ Width ________ Height ________

3. Measure the distance from the camera to the subject.

   Distance from the camera to the subject ________

4. Determine the required width of the horizontal field of view (HFOV). The size of the HFOV depends on the area around the subject matter that you want to include in the photograph. For instance, if you want to photograph a 3-foot-wide (0.9-m-wide) entryway to a building and need to include 5 feet (1.5 m) to either side, then the HFOV is 13 feet (4 m).

   HFOV ________

5. What aspect ratio are you planning to use, 4:3 or 16:9? This will determine the vertical field of view.

   Desired aspect ratio ________

6. Calculate the vertical field of view (VFOV) based on the aspect ratio.

   For 16:9, multiply the HFOV by 0.5625.
   For 4:3, multiply the HFOV by 0.75.

   VFOV ________

7. Is the calculated VFOV going to be high enough? If not, then you may need to adjust the width of the HFOV.
8. What size camera format do you want to use—2/3, ½, 1/3, or ¼ inch?

   Size of camera format ________

9. Determine the width of the camera sensor.

   2/3-inch format = 8.8 mm width
   ½-inch format = 6.4 mm width
   1/3-inch format = 4.8 mm width
   ¼-inch format = 3.2 mm width

   Sensor width ________

10. Calculate the focal length of the lens, using the following formula:

    Focal length = D × Sensor width/HFOV,

    Where D is the distance from the camera to the subject

    Focal length ________

11. Using a light meter, measure the amount of light in the space. If you are using a foot-candle meter, then convert the reading to lux after you are finished.

    Foot-candles ________ Lux ________

12. Does the level of light change throughout the day? If so, be sure you measure the minimum amount of light for a 24-hour period.
13. Assume that facial recognition is going to be required, so you will need 125 pixels per foot. Multiply your HFOV and VFOV by 125.

    HFOV × 125 = ____________ horizontal pixels

    VFOV × 125 = ____________ vertical pixels

14. To obtain the minimum mega-pixel requirement for the digital camera, multiply the horizontal pixels and the vertical pixels together.

    Total number of pixels required = ____________ mega-pixels

15. Based on the total number of pixels required, what size digital camera in mega-pixels is going to be required?

    Digital camera size in mega-pixels _________

16. What type of camera do you want to buy, complementary metal oxide semiconductor (CMOS) or charge-coupled device (CCD)?
17. Based on the above calculations, find a camera on the Internet that will meet the specifications for the above design, and write down the cost of the camera, or the camera and lens if they are purchased separately. Be sure that the camera can operate within the minimum lighting requirements of the space.

    Cost of camera and lens ________

18. Your design is now complete.

FINAL SUMMARY

If you look back on the previous exercise, you will notice that there was no discussion on power supplies, the type of cabling to be used, or storage media. The reason for this is because the more important questions to ask when designing a CCTV system are what is being photographed, how big is the space, what is the level of lighting, and what type of camera and lens is required. Regardless of whether the camera connects through a coaxial cable or CAT-5, or to a DVR or networked-based server, the photographic calculations take precedence over all other decisions.

The majority of CCTV systems now use networked-based cameras and some type of digital storage media. Cameras typically connect through CAT-5 and are powered over PoE switches. For a network-based system, the next consideration is always bandwidth. Depending on the number of cameras in the system and the total number of frames taken per minute, network resources need to be designed appropriately to handle the digital traffic.

When designing a CCTV system, always start with the requirements of the camera specifications, and then work out the connections, storage media, and network resources later.

One final note: Outdoor cameras are typically placed in weatherproof housings that include heaters and cooling fans. Additional power is required when installing these types of enclosures. Always make sure the system power supply can not only power the cameras but also any additional hardware that is required.

FINAL QUESTIONS

1. How does the number of megapixels affect the size of the image?
2. If you increase the focal length of a lens, what happens to the viewing angle? Does it increase or decrease?
3. What does a lower f-stop mean? As an example, what is the difference between a 2.8 f-stop and a 1.2 f-stop?
4. Can you think of ways to adjust the system to help reduce network bandwidth?

APPENDIX

List of Materials and Tools for the Lab Experiments

LIST OF MATERIALS

2011 *National Electrical Code*® book
Safety glasses
Calculator
Ruler
Tape measure
Silicone
Type 66 punch-down block and wall mount
Type 66 punch-down lightning protection block
Cable routing spools for telephone installation
2 telephone handsets

Electronic Parts

8-ohm, 25-watt resistor
Two 47-ohm, ½-watt resistors
Two 100-ohm, ½-watt resistors
Two 330-ohm, ½-watt resistors
Eight 5.6-kiloohm, ½-watt resistors
Twelve 4.7-kiloohm, ½-watt resistors
1-kiloohm, ½-watt resistor
1N4002 diode (with specification sheets)
Red LED (with specification sheets)
Green LED (with specification sheets)
1 double-pull double-throw (DPDT) 12 VDC relay

Hardware

2½-inch D-rings

¾-inch-long mounting screws for the D-rings

Eight 5/16-inch-diameter lag bolts, ¾ inch long

Nine 5/16-inch-diameter bolts, ¾ inch long; lock washers and nuts

1-inch bushing or standard Romex connector

1 metal equipment enclosure (15 in. H × 12 in. W × 4 in. D or something similar in size)

Grounding

Copper busbar for TMGB, having minimum dimensions of 6 mm (0.25 in.) thick × 100 mm (4 in.) wide × 152 mm (6 in.) long

Copper busbar for TGB, having minimum dimensions of 6 mm (0.25 in.) thick × 50 mm (2 in.) wide × 152 mm (6 in.) long

2 busbar mounting kits, each kit including 2 busbar insulators, 2 busbar mounting brackets, and supporting hardware

5/8-inch-diameter copper ground rod, 3 feet long

5/8-inch ground rod clamp

Coaxial cable grounding block

3 ground clamps

Cable

Series-6 coaxial cable

RG-59 coax

Spool of 2-conductor, 18 AWG copper cable, having an unknown length, estimated between 500 and 1,000 feet (you must have access to both ends of the spool)

AWG 6 copper conductor, solid or stranded, bare or green insulated

18 AWG, 2-conductor speaker wire

Computer modem phone cable—preassembled

4-conductor flat telephone cable

4-conductor telephone wire (red, green, yellow, black)

1 roll of 24 AWG, 2-pair cross-connect wire

22 AWG hookup wire for electronic PAD-234 analog digital trainer

2-pair 24 AWG wire

18 AWG, 2-conductor, FPL cable

22 AWG 4-conductor wire

75-ohm outdoor all-weather coaxial cable

10 AWG copper grounding conductor

CAT-5e cable

62.5/125, MM fiber optic cable, with or without a 3-mm (0.1-in.) jacket

Connectors

Male, BNC-type connectors for Series-6 coaxial cable (2-piece or 3-piece)

Male, F-type connectors for RG-59 coaxial cable (crimp style)

Tinned copper, single-hole compression lug for AWG 6 solid or stranded copper conductors

Tinned copper, 2-hole compression lugs for AWG 6 solid or stranded copper conductors

Closed-end crimp connectors

RCA to Banana or BNC (depending on what type of output connector is on the audio generator)

RCA Y adapter for combining stereo to mono

Stereo headphone adapter to RCA cable

RJ-11 (P6C4) modular plug connectors

RJ-11 wall jacks

18 AWG spade connectors

24 AWG spade connectors

RJ-45 (8P8C) modular plugs

2 alligator clips

LC, SC, FC, or ST Leviton Fast Cure fiber connectors

ST Hot Melt® fiber connectors

Fire Alarm System

Silent Knight 5208 fire alarm panel

2-wire smoke detector—ESL 429CT

4-wire smoke detector—ESL 449CT

SPST toggle switch to simulated tamper input

SPST toggle switch to simulated water flow

Pull station—FED 4050-001T-201

2 strobes—WHE RSS24MCW-FR

Audible horn—System Sensor SpectrAlert H12/24 horn

Magnetic door holder—EDW 1508-AQN5

Magnet to test smoke detectors

DPDT 24 VDC power supervision relay (PSR)

Optical Fiber Tools and Supplies

Loctite® 648 adhesive

Loctite® 7649 Primer

Loctite® MSDS sheets

Adhesive bottle needles or adhesive syringe

Fiber safety mat

Fiber trash bin

Fiber-optic tool kit:
- Kevlar shears
- Fiber tweezers
- Fiber jacket stripper
- Fiber buffer stripper
- Fiber crimper
- Fiber scribe
- Multimode 62.5/125 polishing puck
- For LC connectors, Leviton 1.25-mm polishing puck
- Polishing pad

Fiber consumables:
- 99 percent alcohol wipes
- Lint-free wipes
- 0.3-µm film (blue)
- 3-µm film (yellow)
- 12-µm film (dark pink)

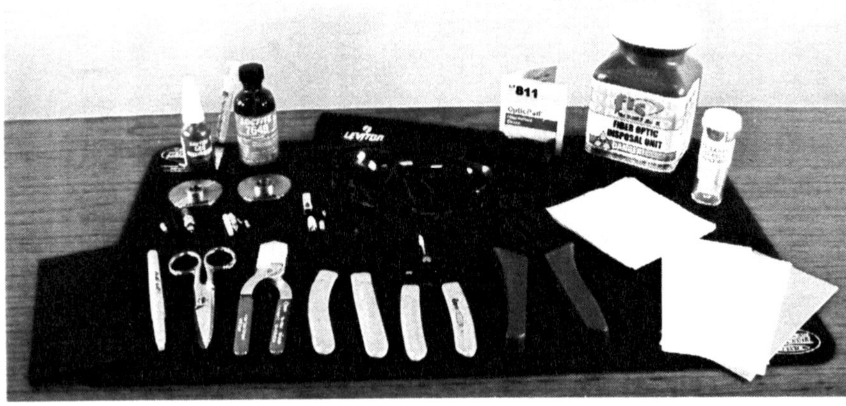

© Cengage Learning 2013

FIGURE APP–1 Fiber-optic tool kit and consumables.

Security System

DSC PC1832 security panel

16.5 V, 40 VA ac power supply

PK 5500 keypad

PIR motion sensor

Acoustic glass break sensor

Acoustic glass break sensor tester

2 sets of door contacts, (NO) closed loop with magnet connected

Bell with steady or yelp option

12-volt, 7-Ahr backup battery

Access Control System

KeriSystems PXL-500 access control panel

12-volt dc, 4-amp power supply for main panel, locks, and siren

Proximity card reader

Wiegand card reader

Proximity cards

Wiegand cards

Electric strike (12-volt dc, fail-secure)

Maglock (12-volt dc)

3 transient protectors

Lighted RTE switch (normally open, momentary switch, 12-volt dc)

Magnetic contacts (normally open, closed-loop when magnet is engaged)

Alarm siren/horn (12-volt dc)

HDTV Digital Antenna

HDTV digital outdoor antenna

Antenna mast, 1½-inch (38.1-mm) pipe

Mast support base for roof mount

Mounting hardware

HDTV with built-in ATSC tuner for HDTV reception

Audio

8-ohm speaker

10-watt audio amplifier

Audio generator

Bogen C20 or C35 amplifier

Shure 522 paging microphone

70-volt or 25-volt, 8-inch (304.2 mm) ceiling speaker, with 13-inch (330.2 mm) grill cover and mounting screws

70-volt or 25-volt transformer volume control

Atlas 81-8R, 24-inch (609.6 mm) ceiling tile bridge (also known as a T-bar) with round hole

12-inch ceiling-mount back box

24-inch × 48-inch (609.6-mm × 1219.2-mm) or 24-inch × 24-inch (69.6-mm × 609.6-mm) ceiling tile

Audio input source: CD or MP3 player

TEST EQUIPMENT

Digital multimeter (DMM)

PAD-234 analog digital trainer

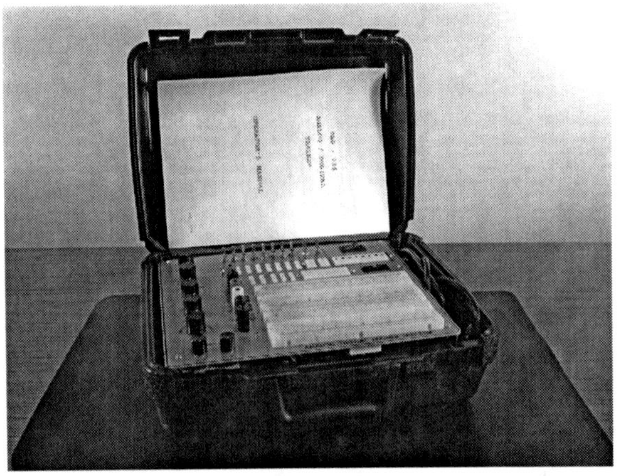

FIGURE APP–2 PAD-234 analog digital trainer.

Radio Shack digital sound level meter

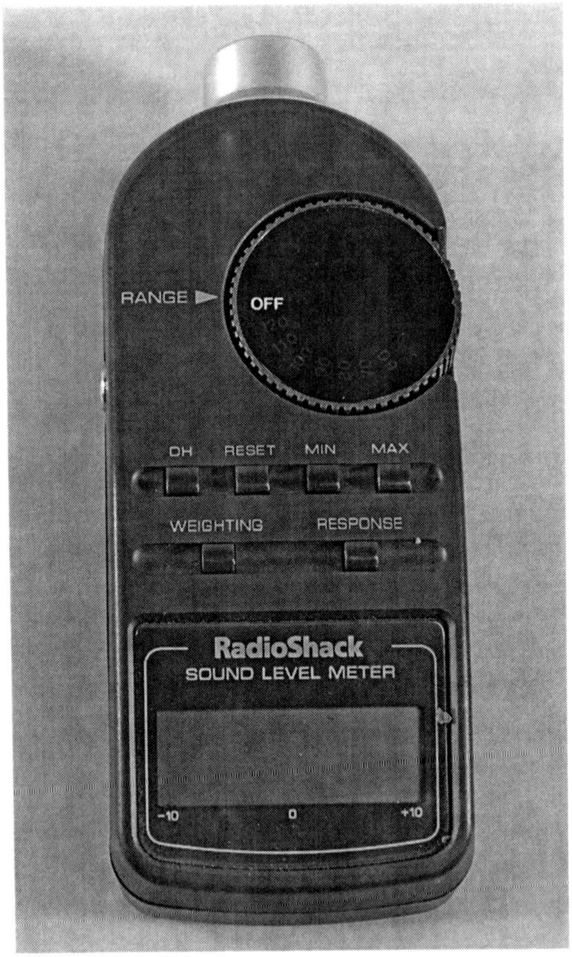

FIGURE APP–3 Sound level meter.

Ideal, LinkMaster Pro XL cable tester

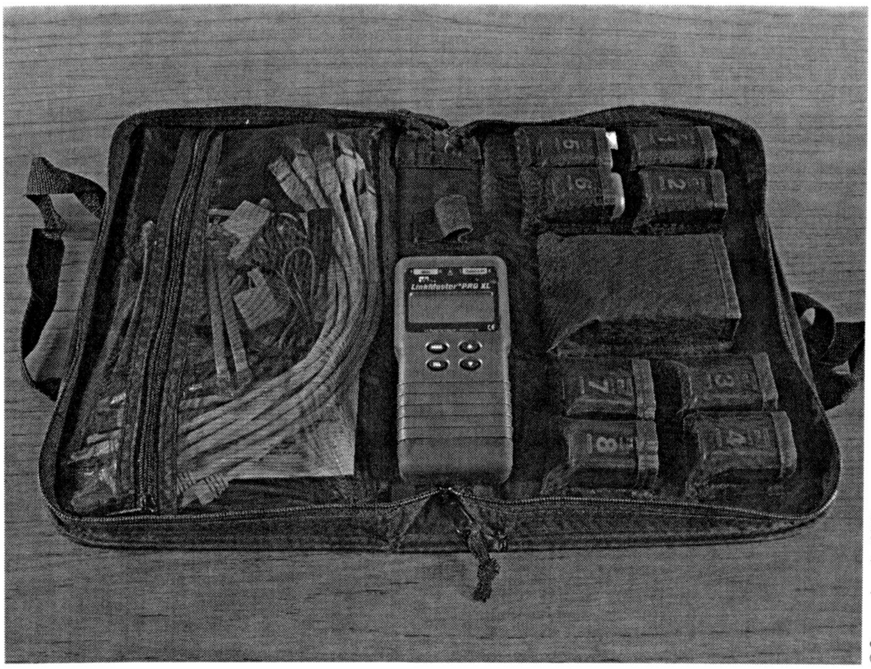

© Cengage Learning 2013

FIGURE APP–4 LinkMaster Pro XL cable tester.

Fiber-optic light source and power meter, with types LC, SC, FC, or ST adaptors

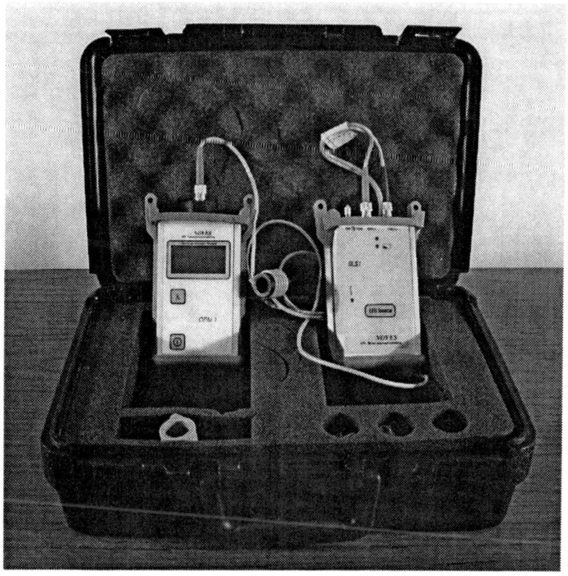

© Cengage Learning 2013

FIGURE APP–5 Fiber-optic light source and power meter.

Fiber-optic microscope (200×)

Hot Melt® oven with type ST connector holder and cooling rack

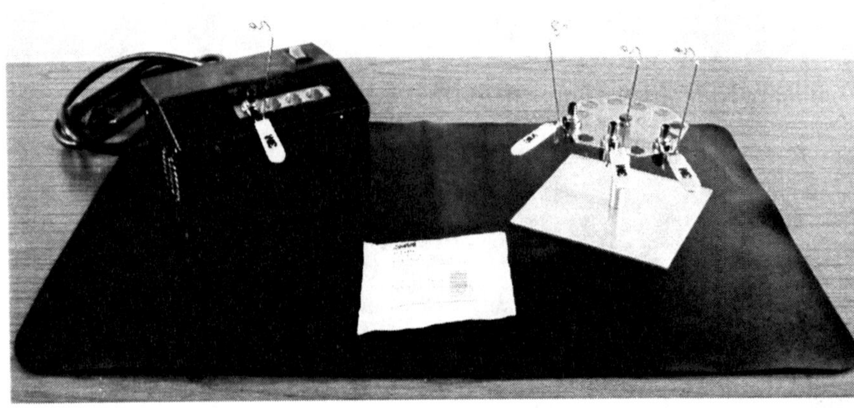

FIGURE APP–6 Hot Melt® oven.

Fluke TS22 test set (butt-set)

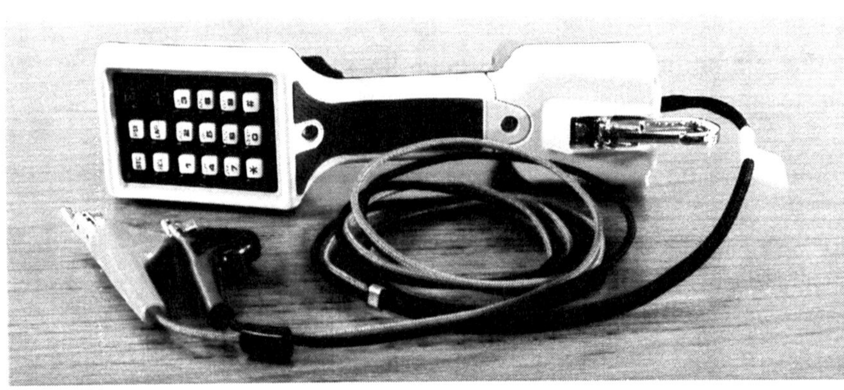

FIGURE APP–7 Fluke TS22 test set.

Ring-It telephone simulator

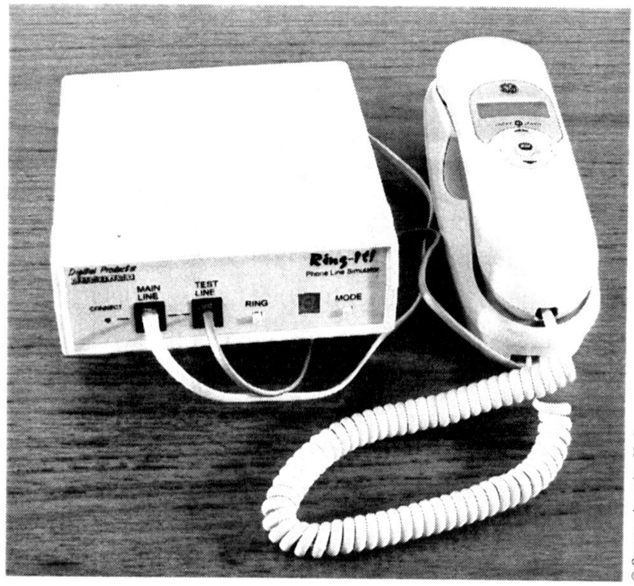

© Cengage Learning 2013

FIGURE APP–8 Ring It telephone simulator.

Antennas Direct SM200 HDTV signal finder
Light meter

TOOL LIST

Screwdrivers
Wire stripper
Coaxial cable stripping tool

© Cengage Learning 2013

FIGURE APP–9 Coaxial cable stripper.

Coaxial cable hex crimping tool for Series-6 and RG-59 cables

FIGURE APP–10 Coaxial cable hex crimping tool.

Compression tool for #6 compression lugs
8P8C, RJ-45 modular plug crimp tool

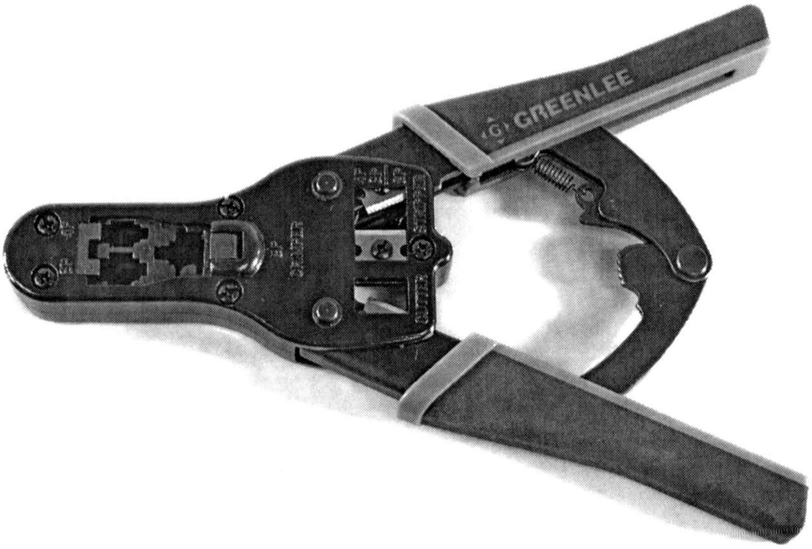

FIGURE APP–11 Modular plug crimping tool.

RJ-11 modular plug stripping and crimping tool

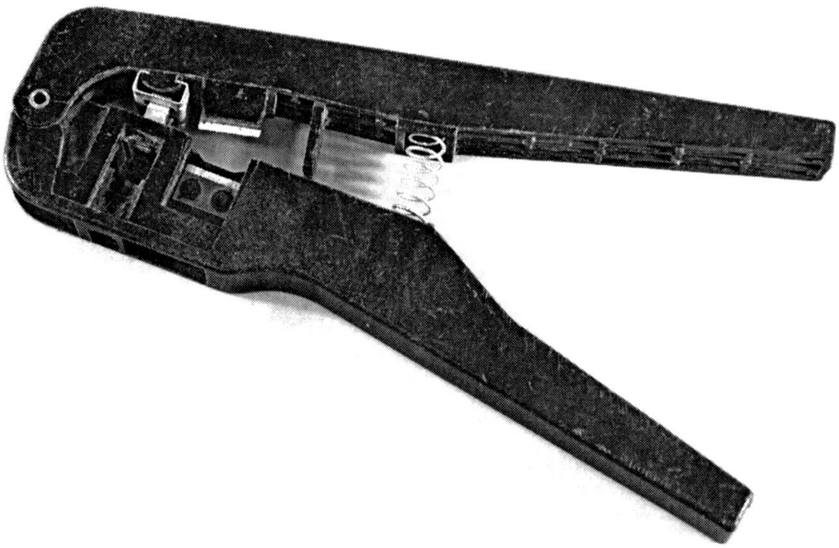

FIGURE APP–12 RJ-11 modular plug stripping and crimping tool.

Drywall or keyhole saw
Crimp tool for closed-end crimp connectors
Crimp tool for 18 and 24 AWG spade connectors

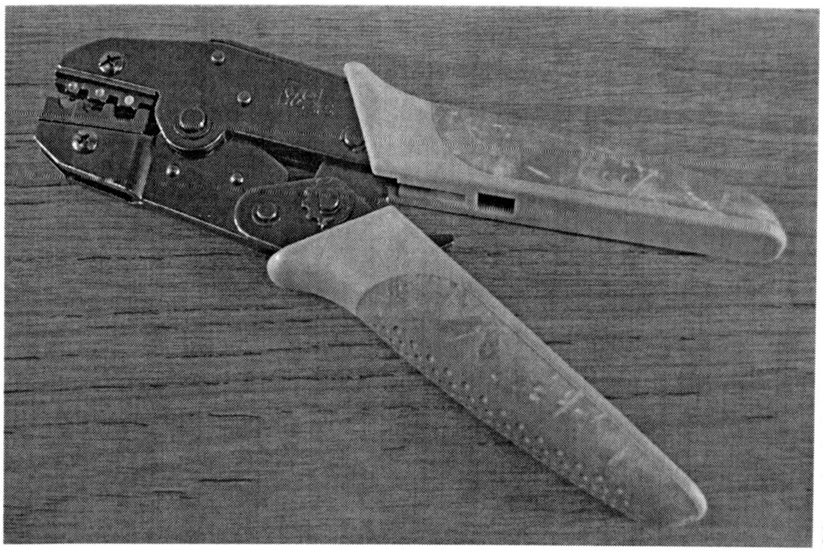

FIGURE APP–13 Crimp tool.

Punch-down tool for 66-block

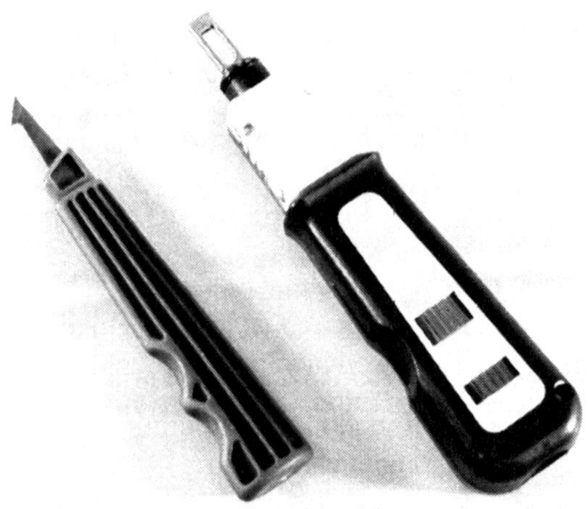

FIGURE APP–14 Type 66 punch-down tool.

Blades for type 66 punch-down tool

Spudger

NOTES

NOTES

NOTES

NOTES

NOTES

NOTES

CPSIA information can be obtained
at www.ICGtesting.com
Printed in the USA
FFOW01n1733290618
47278667-50202FF